职业教育大数据技术专业"互联网+"创新教材

大数据平台部署与运维

主　编　王安曼　章增优
副主编　徐欣欣　宁海元
参　编　张学清　郑定超

机械工业出版社

本书以大数据运维工程师为职业培养方向，以职业岗位的"典型工作过程"为导向，融入行动导向教学法，将教学内容与职业能力相对接、单元项目与工作任务相对接。本书以实训为主体，配合实训指导视频，结合适当的知识讲解，让大数据技术学习者能快速掌握大数据平台部署与运维相关的实操能力，并建立基础理论认知。本书共 8 个项目，内容包括认识大数据、配置平台基础环境、部署 Hadoop 框架、使用 HDFS、MapReduce 编程、部署与使用 HBase、部署与使用 Hive、部署与使用 Spark，涵盖了当前大数据技术领域的主要技术。

本书可作为各类职业院校大数据技术等相关专业的教学用书，也可作为从事大数据平台运维相关工作者的参考用书。

本书配有理论学习微课、任务实践演示微课（扫描书中二维码观看），还配有电子课件等资源，选用本书作为授课教材的教师可以从机械工业出版社教育服务网（www.cmpedu.com）免费注册后进行下载或联系编辑（010-88379807）咨询。

图书在版编目（CIP）数据

大数据平台部署与运维/王安曼，章增优主编. —北京：机械工业出版社，2023.4（2024.12重印）
职业教育大数据技术专业"互联网+"创新教材
ISBN 978-7-111-72840-5

Ⅰ．①大…　Ⅱ．①王…　②章…　Ⅲ．①数据处理-职业教育-教材　Ⅳ．①TP274

中国国家版本馆CIP数据核字（2023）第047628号

机械工业出版社（北京市百万庄大街22号　邮政编码100037）
策划编辑：张星瑶　　　　　责任编辑：张星瑶
责任校对：李小宝　王　延　封面设计：马若濛
责任印制：单爱军
北京虎彩文化传播有限公司印刷
2024 年 12 月第 1 版第 3 次印刷
210mm×285mm・12印张・304千字
标准书号：ISBN 978-7-111-72840-5
定价：45.00元

电话服务　　　　　　　　　网络服务
客服电话：010-88361066　　机　工　官　网：www.cmpbook.com
　　　　　010-88379833　　机　工　官　博：weibo.com/cmp1952
　　　　　010-68326294　　金　书　网：www.golden-book.com
封底无防伪标均为盗版　机工教育服务网：www.cmpedu.com

前言

《国家职业教育改革实施方案》中提出要落实立德树人根本任务,深化专业、课程、教材改革,提升实习实训水平,努力实现职业技能和职业精神培养高度融合。在这样的背景下,本书根据课程标准的要求,以大数据运维工程师作为岗位培养方向,以"岗位能力"作为培养目标,以"典型工作场景"作为项目背景,将"典型工作任务"作为课程主要内容,引导学生在"理实一体"的学习过程中,掌握岗位能力要求。

主要内容:

项目1——认识大数据。了解大数据的概念、特征、应用场景,并在认识大数据项目实施流程的基础上,了解大数据运维工程师的岗位能力要求,并作为本书学习的能力目标。

项目2——配置平台基础环境。学习Linux操作系统简介与Linux常用指令,完成三个典型工作任务:安装操作系统、配置静态IP、远程登录,实现大数据集群的虚拟化环境部署与访问。

项目3——部署Hadoop框架。学习Hadoop相关基础概念,在了解Hadoop部署模式基础上,由易到难,依次完成三个任务:部署单机模式Hadoop、部署伪分布模式Hadoop、部署全分布模式Hadoop,最终实现全分布模式Hadoop的部署工作。

项目4——使用HDFS。学习HDFS相关基础概念,完成两个任务:使用HDFS的Web界面监管HDFS集群状态、使用Shell管理HDFS文件与目录。在此基础上,进一步学习HDFS的体系结构与运行机制。

项目5——MapReduce编程。学习MapReduce相关基础概念,理解MapReduce核心思想、设计理念、编程思想,在认识MapReduce词频统计的示例基础上,完成任务:利用MapReduce编程实现词频统计功能。进一步学习MapReduce运行机制。

项目6——部署与使用HBase。学习HBase相关基础概念,认识HBase数据模型与系统架构,完成三个任务:部署伪分布式HBase、部署完全分布式HBase、利用Shell操作HBase。进一步学习HBase的Region服务器、HLog、Store工作原理。

项目7——部署与使用Hive。学习Hive相关基础概念,区别Hive与传统数据库的关系,完成四个任务:部署本地模式Hive、部署远程模式Hive、利用Hive实现数据导入、利用Hive实现词频统计。进一步学习Hive的架构、运行机制等。

项目8——部署与使用Spark。学习Spark相关基础概念,了解Spark与Hadoop的关系,认识Spark的部署形式及Shell操作,完成四个任务:部署单机模式Spark、部署Spark集群、使用Spark Shell编写代码、使用Scala编写Spark程序。进一步学习Spark的架构与运行机制、了解RDD的设计与原理。

本书特点:

本书设计过程中,有机融入习近平新时代中国特色社会主义思想、社会主义核心价值观、中华优秀传统文化、职业理想和职业道德等内容。在案例实践操作中注重知行合一,增强学生勇于探索的创新精神、善于解决问题的实践能力,培养学生精益求精的大国工匠精神,不断提升学生的课程学习体验、学习效果,从而达到价值塑造、知识传授和能力培养三者融为一体。

本书为融媒体新形态教材,以活页式的设计理念,灵活跟进教学实践,支持多种形式重构:以项目为基点,选择合适的项目组合教学;以任务为单位,选择符合学生能力、学习目标的任务内容开展实训;以理论、实训为区分,将整本书分为理论部分与实训部分,分别开展教学。还对应配套理论学习微课、任务实践演示微课、和其他电子教学资源。

本书设有笔记边栏部分,侧重实训教学,设计为"理实一体"的体例形式,左边展示操作过程,右边同步展示实训要点、理论知识点、特别说明等,在实践过程中,适时地融入理论指导,帮助读者加深对理论的理解,并更好地完成实训任务。

教学建议:

本书对应大数据平台部署与运维课程,建议课程安排为64学时,其中,动手操作为42学时,理论学习为22学时,强调"做中学",通过动手实践的方式,强化对知识的理解与记忆。具体学时安排如下:

教学安排建议

项　　目	动手操作学时	理论学习学时	合　　计
项目1　认识大数据	0	2	2
项目2　配置平台基础环境	4	2	6
项目3　部署Hadoop框架	10	2	12
项目4　使用HDFS	2	2	4
项目5　MapReduce编程	2	2	4
项目6　部署与使用HBase	8	4	12
项目7　部署与使用Hive	8	4	12
项目8　部署与使用Spark	8	4	12
总　　计	42	22	64

学习本书内容前,应具备一定的Linux操作系统知识,并能熟练使用Linux常用指令,若学校未开设相关前序课程,建议适当增加项目2的学时。

编写队伍:

本书由浙江工贸职业技术学院王安曼、章增优任主编,浙江工贸职业技术学院徐欣欣、杭州玳数科技有限公司宁海元任副主编,参与编写的还有浙江工贸职业技术学院张学清和浙江东方职业技术学院郑定超。

本书由宁海元负责分析岗位典型工作任务,提供企业项目案例,并提供内容合理化参考意见;由王安曼、章增优负责大纲设计与体例设计;王安曼编写项目1~5,章增优、王安曼共同编写项目6,徐欣欣编写项目7、项目8;全体成员参与微课、电子课件、教案、题库等资源的制作。

本书是在2019年浙江省教育厅一般科研项目《基于大数据基础实验平台的活页教材建设》(Y201942899)的背景下,策划并编写完成的。

由于作者水平有限,书中难免存在错误或不妥之处,恳请读者批评指正,编者邮箱:wanganman@zjitc.edu.cn。

编　者

二维码索引

序号	视频名称	二维码	页码	序号	视频名称	二维码	页码
1	项目1 知识准备		1	11	项目2 任务2 配置静态IP		23
2	项目1 拓展学习		4	12	项目2 任务3 远程登录		26
3	项目2 知识准备 Linux操作系统		8	13	项目2 拓展学习		31
4	项目2 知识准备 常用指令 文件与目录管理1		8	14	项目3 知识准备		34
5	项目2 知识准备 常用指令 文件与目录管理2		8	15	项目3 任务1 部署单机模式Hadoop		34
6	项目2 知识准备 常用指令 文本编辑		10	16	项目3 任务2 部署伪分布模式Hadoop		41
7	项目2 知识准备 常用指令 用户管理		10	17	项目3 任务3 部署全分布模式Hadoop1		49
8	项目2 知识准备 常用指令 文件权限管理		11	18	项目3 任务3 部署全分布模式Hadoop2		55
9	项目2 知识准备 常用指令 系统管理		12	19	项目3 任务3 部署全分布模式Hadoop3		60
10	项目2 任务1 安装操作系统		13	20	项目3 拓展学习		62

（续）

序号	视频名称	二维码	页码	序号	视频名称	二维码	页码
21	项目4 知识准备		70	32	项目6 任务1 部署伪分布式HBase		103
22	项目4 知识准备 Shell命令1		71	33	项目6 任务2 部署完全分布式HBase		107
23	项目4 知识准备 Shell命令2		71	34	项目6 任务3 利用Shell操作HBase		113
24	项目4 任务1 使用HDFS的Web界面		71	35	项目6 拓展学习		118
25	项目4 任务2 使用Shell管理HDFS文件与目录		75	36	项目7 知识准备		124
26	项目4 拓展学习		80	37	项目7 任务1 部署本地模式Hive1		126
27	项目5 知识准备		84	38	项目7 任务1 部署本地模式Hive2		128
28	项目5 任务 MapReduce编程实现词频统计		87	39	项目7 任务1 部署本地模式Hive3		131
29	项目5 拓展学习		94	40	项目7 任务2 部署远程模式Hive		135
30	项目6 知识准备		100	41	项目7 任务3 利用Hive实现数据导入		143
31	项目6 知识准备 Shell操作		102	42	项目7 任务4 利用Hive实现词频统计		151

（续）

序号	视频名称	二维码	页码	序号	视频名称	二维码	页码
43	项目7 拓展学习		155	47	项目8 任务2 部署Spark集群		165
44	项目8 知识准备		160	48	项目8 任务3 使用Spark Shell编写代码		168
45	项目8 知识准备 Shell		161	49	项目8 任务4 使用Scala编写Spark程序		172
46	项目8 任务1 部署单机模式Spark		162	50	项目8 拓展学习		177

目录

- 前言
- 二维码索引
- 项目1　认识大数据 .... // 1
  - 拓展学习 .... // 4
  - 项目小结 .... // 5
- 项目2　配置平台基础环境 .... // 7
  - 任务1　安装操作系统 .... // 13
  - 任务2　配置静态IP .... // 23
  - 任务3　远程登录 .... // 26
  - 拓展学习 .... // 31
  - 项目小结 .... // 31
  - 实战强化 .... // 31
- 项目3　部署Hadoop框架 .... // 33
  - 任务1　部署单机模式Hadoop .... // 34
  - 任务2　部署伪分布模式Hadoop .... // 41
  - 任务3　部署全分布模式Hadoop .... // 49
  - 拓展学习 .... // 62
  - 项目小结 .... // 66
  - 实战强化 .... // 66
- 项目4　使用HDFS .... // 69
  - 任务1　使用HDFS的Web界面 .... // 71
  - 任务2　使用Shell管理HDFS文件与目录 .... // 75
  - 拓展学习 .... // 80
  - 项目小结 .... // 82
  - 实战强化 .... // 82
- 项目5　MapReduce编程 .... // 83
  - 任务　MapReduce编程实现词频统计 .... // 87
  - 拓展学习 .... // 94
  - 项目小结 .... // 97
  - 实战强化 .... // 97
- 项目6　部署与使用HBase .... // 99
  - 任务1　部署伪分布式HBase .... // 103

任务2	部署完全分布式HBase	// 107
任务3	利用Shell操作HBase	// 113
拓展学习		// 118
项目小结		// 121
实战强化		// 122

项目7　部署与使用Hive .... // 123

任务1	部署本地模式Hive	// 126
任务2	部署远程模式Hive	// 135
任务3	利用Hive实现数据导入	// 143
任务4	利用Hive实现词频统计	// 151
拓展学习		// 155
项目小结		// 156
实战强化		// 157

项目8　部署与使用Spark .... // 159

任务1	部署单机模式Spark	// 162
任务2	部署Spark集群	// 165
任务3	使用Spark Shell编写代码	// 168
任务4	使用Scala编写Spark程序	// 172
拓展学习		// 177
项目小结		// 179
实战强化		// 180

参考文献 .... // 181

项目1 认识大数据

项目概述

A公司是一家大数据产品研发企业，公司根据客户的大数据业务场景需求，开发完成各类大数据平台项目。大学毕业生小王新入职公司，任职大数据运维工程师，同事称呼他为王工。公司为新员工制定了员工培养方案，帮助员工认识大数据、了解岗位职责等。

本项目作为职场入门第一课，通过介绍大数据行业背景、大数据项目实施流程、大数据运维工程师岗位，让大数据职场新人快速进入岗位角色。

学习目标

1. 了解大数据概念、特征、应用场景。
2. 了解大数据项目实施流程。
3. 了解大数据运维工程师工作职责、能力素养要求。

思维导图

项目思维导图如图1-1所示。

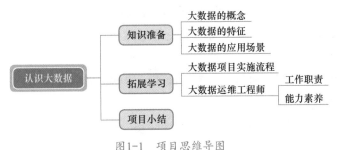

图1-1 项目思维导图

扫码看视频

知识准备

1. 大数据的概念

什么是大数据？大数据并不是指数据量大于某个阈值，而是数据集超出了传统数据库处理能力。麦肯锡全球研究所给出的定义是：一种规模大到在获取、存储、管理、分析方面大大超出了传统数据库软件工具能力范围的数据集合，具有海量的数据规模、快速的数据流转、多样的数据类型和价值密度低四大特征。

大数据技术的战略意义不在于掌握庞大的数据信息，而在于对这些含有意义的数据进行专业化处理。换而言之，如果把大数据比作一种产业，那么这种产业实现盈利的关键在于提高对数据的"加工能力"，通过"加工"实现数据的"增值"。

大数据必然无法用单台的计算机进行处理，必须采用分布式架构。它的特色在于对海量数据进行分布式数据挖掘。因此，大数据必须依托云计算的分布式处理、分布式数据库和云存储、虚拟化技术，两者密不可分。

2. 大数据的特征

大数据具有4V特征：大量（Volume）、多样（Variety）、高速（Velocity）、价值（Value），如图1-2所示。

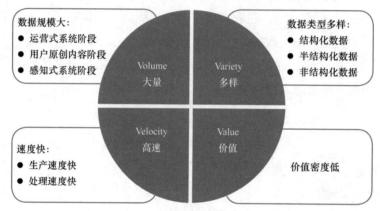

图1-2　大数据4V特征

（1）大量（Volume）

数据规模大是大数据的首要特征。人类社会从数字化技术诞生以来，数据生产方式经历了三个阶段：

1）运营式系统阶段，数据仅有少数运营人员生产并发布。

2）用户原创内容阶段，即自媒体时代，全民都是数据的生产者与发布者，数据生产速度大幅提升。

3）感知式系统阶段，即物联网时代，各类摄像头、传感器实时生产数据、记录数据，其数据生产速度较人工数据再一次发生质的跨越。

当前正处于感知式系统阶段，数据正呈现爆发性增长，相关计算单位见表1-1。

表1-1　存储单位换算

单　位	换算公式
Byte（字节）	1Byte=8bit
KB（KiloByte，千字节）	1KB=1024Byte
MB（MegeByte，兆字节）	1MB=1024KB
GB（GigaByte，吉字节）	1GB=1024MB
TB（TeraByte，太字节）	1TB=1024GB
PB（PetaByte，拍字节）	1PB=1024TB
EB（ExaByte，艾字节）	1EB=1024PB
ZB（ZettaByte，泽字节）	1ZB=1024EB

（2）多样（Variety）

大数据的数据来源众多，决定了数据类型的多样性。按照数据类型可将大数据分为三类：结构化数据、半结构化数据、非结构化数据。

1）结构化数据：是由二维表结构来逻辑表达和实现的数据，严格遵循数据格式与长度规范，可以存储在传统关系型数据库中，如财务系统、医疗系统、信息管理系统等，数据间具有因果关系。

2）半结构化数据：相对于结构化数据，其数据结构变化很大，因果关系弱。通常采用HTML文档、邮件、网页等形式表达。如员工简历信息，不是所有的员工简历都具有相同的字段，具有一定的灵活性。

3）非结构化数据：指图像、视频、音频文件，数据间没有因果关系。

（3）高速（Velocity）

从两个角度理解大数据的高速：

1）数据生产速度快：在物联网时代，基于传感器等数据生产方式的多样性与高速网络带宽的共同作用下，实现了高效的数据生产。

2）数据处理速度快：大数据时代的应用要求大数据具有实时分析结果的能力，指导生活和生产实践。例如，当前各大APP基于用户行为实现的用户内容精准推送，在极大程度上决定了其在同类市场中的用户占有率。

（4）价值（Value）

大数据的价值密度远远低于传统关系型数据库中的数据。例如，小区监控视频24小时记录小区监控画面，需要消耗大量存储空间，却仅在特殊情况下（如盗窃案）的一小段视频才有意义。

但是任何价值都需要以海量数据作为基础，否则大数据技术将无从谈起。从价值角度来说，当前大数据技术需要解决如何通过强大的算法迅速挖掘出海量数据中的价值。

3. 大数据的应用场景

大数据技术在各行各业中的应用见表1-2。

表1-2 大数据的应用场景

行 业	应 用
金融	麦肯锡研究指出，金融业在大数据价值潜力指数中排第一。例如： 1）基于用户行为大数据，构建用户画像，实现个性化智慧营销、服务创新等 2）基于内部多元异构数据和外部征信数据，优化风控管理 3）通过交叉领域的数据共享分析，优化产品创新
医疗	1）基于海量的病例、报告、治愈方案、药物方案等，优化医疗方案，提供最佳治疗方法 2）监控居民人体数据，早治疗、早康复，有效预防、预测疾病
零售	1）通过对区域人口、消费、习惯、产品认知等大数据分析，得到更精准的市场定位 2）分析消费者的消费行为、价值取向等，挖掘零售市场新需求 3）根据区域用户购物行为，优化物流仓储网络，提高物流效率，降低物流成本
娱乐	根据用户行为，推送符合用户喜好的影视、文本、广告等内容
城市管理	1）利用道路监控系统及车载GPS大数据，实现实时道路路况监控等指挥交通业务 2）基于各类传感器，实现景区、城市环保监测及决策

拓展学习

1. 大数据项目实施流程

大数据项目实施有其独有的流程，可以大致分为六个阶段：项目规划、数据治理、项目设计、项目应用、迭代与实施、应用推广。大数据项目实施的流程如图1-3所示。

扫码看视频

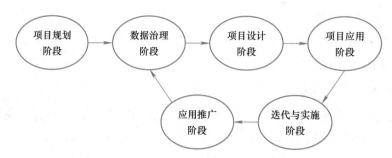

图1-3　大数据项目实施流程

（1）项目规划阶段

该阶段为前期准备工作，了解客户的实际情况、需求、业务模式和现有数据情况等，然后完成数据架构规划。

（2）数据治理阶段

该阶段进行数据采集和数据预处理，获得高质量的数据，数据到位才能保障项目的有效推进和执行。

（3）项目设计阶段

该阶段根据具体数据需求对数据的使用模式及数据特点来设计逻辑数据模型。

（4）项目应用阶段

该阶段根据逻辑数据模型进行设计，从物理存储、文件系统、物理数据库到大数据平台进行技术选型，形成具体的功能规划。

（5）迭代与实施阶段

该阶段对大数据项目进行不断验证、修正、实施，然后根据客户试用评估结果扩展大数据的应用范围。

（6）应用推广阶段

该阶段进行有效的实施推广，让数据"活"起来，源源不断产生价值。

2. 大数据运维工程师

随着大数据产业进入高速发展期，大数据运维工程师已成为近年来的热门职业之一。

（1）工作职责

大数据运维工程师的具体工作职责如下。

1）负责Hadoop、HBase、Spark、Kafka、Redis等大数据生态圈的组建。

2）负责处理大数据集群的各类异常和故障，并能区分故障等级，优先处理影响实时性业务的故障。

3）以可控的方式，高效地完成变更工作，包括配置管理和发布管理。

4）负责大数据平台的容量规划、扩容。

5）负责对大数据平台的定期检查，进行性能分析与调优。

6）协助优化大数据平台架构，支持平台能力和产品不断迭代。

（2）能力素养

大数据运维工程师的能力素养见表1-3。

表1-3 大数据运维工程师的能力素养

能力素养	具体要求
部署大数据平台	掌握Linux操作系统的基础知识
	掌握Hadoop、Spark等集群基础知识
	掌握数据库理论的基础知识
	掌握计算机网络理论的基础知识
	能够安装部署大数据系统，能够对网络、软硬件进行配置
监控大数据平台	能够安装和使用集群监控工具，监控大数据平台运行情况
	能够对大数据平台的故障进行定位、分析和解决
	能够将维护日志、故障、问题记录等形成运维报告
优化大数据平台	能够对运维报告进行分析，根据分析结果对软硬件环境及运维工作机制进行调优

项目小结

本项目作为大数据运维工程师的"新人入职培训"项目，介绍了大数据概念、特征、应用场景，阐述了大数据运维工程师工作职责、能力素养要求，讲解了大数据项目实施流程。

项目 2 配置平台基础环境

●项目概述

A公司根据客户的大数据业务场景需求，开发完成各类大数据平台项目。由于大数据项目通常运行于Linux操作系统环境中，因此，公司要求大数据运维工程师王工为客户的服务器安装Linux操作系统，并对Linux操作系统配置基础网络环境，实现远程访问。

本项目针对"配置平台基础环境"的典型工作任务，根据任务的先后顺序，将其分为三个子任务：安装操作系统、配置静态IP、远程登录。

在开展任务前，需要掌握必要的理论知识：Linux操作系统简介与Linux常用指令。指令包括：文件与目录管理、文本编辑、用户管理、文件权限管理、系统管理。熟练掌握Linux常用指令是大数据运维工程师及大数据实施工程师的基本技能。

完成任务后，进一步了解拓展知识——国产Linux操作系统。

●学习目标

1. 熟悉Linux操作系统，了解其在服务器领域的重要性及优势。
2. 熟练掌握Linux常用指令，包括文件与目录管理、文本编辑、用户管理、文件权限管理、系统管理。
3. 掌握Linux操作系统的安装。
4. 熟练掌握静态IP设置。
5. 熟练掌握SSH远程登录。

●思维导图

项目思维导图如图2-1所示。

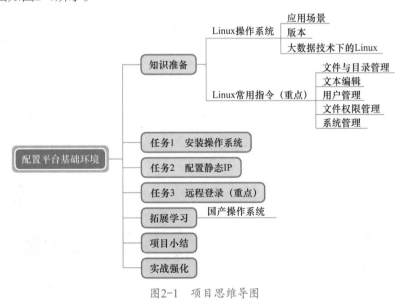

图2-1 项目思维导图

知识准备

1. Linux操作系统

Linux是一套开源、免费、自由传播的类似UNIX的操作系统,最初是由芬兰赫尔辛基大学学生Linus Torvalds于1991年在网络上发布的。

扫码看视频

(1)应用场景

Linux操作系统具有开放性、多任务、多用户的特点,具有良好的用户界面,设备独立,具有丰富的网络功能、可靠的安全性和良好的可移植性,再加上免费和源代码开放,因此其应用领域越来越广泛,主要应用在以下三个场景:IT服务器、嵌入式系统、个人桌面系统,见表2-1。

表2-1 Linux操作系统的主要应用场景

场 景	介 绍
IT服务器	Linux操作系统作为企业级服务器的应用十分广泛,利用Linux操作系统可以为企业构架WWW服务器、数据库服务器、负载均衡服务器、邮件服务器、域名服务器(Domain Name Server,DNS)、代理服务器(透明网关)、路由器等
嵌入式系统	从互联网设备(路由器、交换机、防火墙、负载均衡器等)到专用的控制系统,如自动售货机、手机、个人数字助理(PDA)、家用电器等
个人桌面系统	Linux操作系统完全可以满足日常的办公及家用需求,如浏览器、办公室软件、电子邮件、实时通信、多媒体应用等

(2)版本

Linux版本分为两类:内核(Kernel)版本和发行(Distribution)版本。

内核,指的是控制计算机硬件工作的软件集合,主要模块包括:存储管理、CPU和进程管理、文件系统、设备管理和驱动、网络通信,以及系统的初始化引导、系统调用等。内核版本是由Linus Torvalds领导下的开发小组开发出来的系统内核版本号。

发行版本,指的是一些组织或公司将Linux内核与应用软件和文档组合起来,并提供一些安装界面和系统设置与管理工具。常见的发行版本包括:Slackware、Redhat、Debian、Fedora、TurboLinux、SUSE、CentOS、Ubuntu和国产的中科红旗、麒麟等。其中,CentOS(Community Enterprise Operating System,社区企业操作系统)是目前国内互联网公司使用最多的Linux发行版本。

(3)大数据技术下的Linux

Hadoop作为大数据主要技术框架,可以运行在Linux、Windows和一些类UNIX操作系统上,但Hadoop官方真正支持的操作系统是Linux。因此在Linux操作系统上安装运行Hadoop是首选方案。

2. Linux常用指令

(1)文件与目录管理

文件与目录管理的相关命令是Linux常用命令的基础,是系统运维工程师的基本能力要求。常见的文件与目录管理命令有pwd、cd、ls、mkdir、cp、mv、rm、find、tar,见表2-2。

扫码看视频　　　　扫码看视频

表2-2 文件与目录管理常用指令

指令	指令介绍		
pwd	print working directory，打印当前工作目录的绝对路径		
	格式	pwd	
cd	change directory，更改当前工作目录到目标目录，路径的表达方式可以是相对路径或绝对路径		
	格式	cd <路径>	
	示例	cd /data #进入根路径下的data目录 cd data #进入当前路径下的data目录	
ls	list，列出路径下的文件列表		
	格式	ls [选项] [文件或目录] -l：以详细信息等形式展示出当前目录下的文件 -a：显示当前目录下的所有文件（包括隐藏文件）	
mkdir	make directory，创建目录，需要保证新的目录不与同路径下的目录重名		
	格式	mkdir [选项] 目录名 -p：当参数为路径时，默认仅创建最后一级目录，使用-p参数，则会同时创建路径中不存在的目录，以通过一条命令创建多级目录	
	示例	mkdir /data/dir1 #在/data文件夹中创建dir1文件夹，要求/data文件夹已存在 mkdir -pa/b #在当前路径下创建a文件夹，同时在a文件夹下创建b文件夹	
cp	copy，将一个或多个源文件复制到指定的目录		
	格式	cp [选项] 源文件或目录 目的目录 -R：递归处理，将指定目录下的文件及子文件一并处理	
	示例	cp 1.xml /data #将当前路径下的1.xml文件复制到/data文件夹中	
mv	move，移动文件或目录		
	格式	mv 源文件或目录 目标目录 1. 如果同时指定两个及以上的文件或目录，且最后一个目的地是一个已经存在的目录，则该指令会将前面所有文件与目录移动到最后一个目录中 2. 如果操作对象是在同一目录下的两个文件，则其功能为修改文件名	
	示例	mv 1.xml /data #将当前路径下的1.xml文件移动到/data文件夹中 mv 1.xml 2.xml #将当前路径下的1.xml文件重命名为2.xml	
rm	remove，删除目录中的文件或目录		
	格式	rm [选项] 文件或目录 -r：删除目录及其子目录下所有文件 -f：强制删除文件或目录，不需要一一确认	
	示例	rm -rf /data #强制删除/data目录及其子目录下所有文件	
find	在指定目录下查找文件		
	格式	find 搜索路径 [选项] 搜索关键词 关键词可以是文件名、文件大小、文件所有者等 -name：根据文件名查找 -size：根据文件大小查找	
	示例	find / -name *.xml #查找/目录下所有文件名以.xml结尾的文件	

（续）

指令	指令介绍	
tar	打包多个目录或文件，与压缩命令一起使用，实现多文档的打包并压缩；也可以将压缩包中的文件和目录释放出来	
	格式	tar [选项] 目录 -c：产生.tar打包文件 -C <目录>：切换工作目录，先进入指定目录再执行压缩或解压缩操作 例如：1）仅压缩特定目录中的内容；2）解压缩到特定目录（更常用） -v：打包时显示详细信息 -f <文件名>：指定压缩后的文件名 -z：打包，同时通过gzip指令压缩备份文件，压缩后的格式为.tar.gz -x：从打包文件中还原文件
	示例	tar -zcf log.tar.gz 1.log #将1.log文件打包并以gzip格式压缩，生成文件名为log.tar.gz tar -zxf log.tar.gz -C /data #将log.tar.gz解压缩至/data路径下

（2）文本编辑

文本编辑使用vi与vim命令，两个命令使用方式相似，编辑器有三种模式，模式转换关系与各模式下主要功能如图2-2所示，常用操作见表2-3。

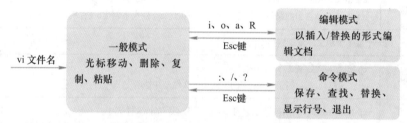

图2-2　vi编辑器三种模式的模式转换关系与各模式下主要功能

表2-3　vi编辑器常用操作方法

按键	作用
i	在指针当前位置进入编辑模式
o	在指针当前位置的下一行插入新的一行，并进入编辑模式
/word	在指针之后查找一个字符串word，按<n>键向后继续搜索
:w	保存文本
:q	退出vim
:q!	强制退出，所有改动不生效
:set nu	显示行号

（3）用户管理

Linux是一个多用户、多任务的操作系统，在同一台Linux主机中，可以同时登录多名用户。每一个系统使用者必须以一个账户的身份登录系统，系统根据不同账号为用户提供不同工作环境。UID是系统中的用户标识符，每个用户具有唯一的UID。

为了实现用户权限的高效管理，Linux引入"用户组"概念，对同一个用户组内的多名用户统一分配权限。GID是系统中的组标识符,每个组具有唯一的GID。

用户管理常用指令见表2-4。

表2-4 用户管理常用指令

指令	指令介绍	
useradd	创建一个新账号	
	格式	useradd [选项] 用户名
	示例	useradd hadoop #创建hadoop用户
passwd	为用户设置登录密码 root用户可以设置所有人密码,普通用户只能修改自己的密码	
	格式	passwd [选项] 用户名
	示例	passwd hadoop #为hadoop用户设置密码
su	switch user,切换用户	
	格式	su [选项] 用户名 -l:切换用户的同时,切换到用户的Home目录,同时更新用户的环境变量 选项形式可以是"-l"或"-"
	示例	su - root #切换到root用户,切换过程需要输入root用户密码
sudo	switch user do,切换用户,让普通用户执行需要特殊权限的指令 使用sudo指令之前,需要使用visudo指令先配置/etc/sudoers文件。通过在sudoers文件中逐条添加配置信息,可以实现为特定用户提升账号权限。完成配置信息后,使用者可以在登录普通用户的情况下,使用sudo指令临时切换为root用户执行普通用户无权限执行的指令	
	格式	sudo [指令]
	示例	sudo reboot #让普通用户执行重启指令,执行过程需要输入用户的密码

(4)文件权限管理

通过指令ls -l可以查看当前路径下每个文件的属性信息,如图2-3所示,其中,部分文件属性说明见表2-5。

扫码看视频

```
     文件类型        文件链接数  文件所属组      文件最近修改时间
drwxr-xr-x. 2 root root     6 3月   6 09:24 20200306
-rw-------. 1 root root  1759 2月  11 16:45 anaconda-ks.cfg
-rw-r--r--. 1 root root  1790 2月  11 17:00 initial-setup-ks.cfg
-rw-r--r--. 1 root root     0 3月  13 23:04 test
   文件权限      文件所有者   文件大小                  文件名
                            (单位:B)
```

图2-3 ls -l查看文件属性

表2-5 文件属性说明

属性栏	说明
文件类型	Linux文件类型包括: 1. 普通文件,用"-"表示,分为二进制文件与文本文件 2. 目录文件,用"d"表示,用于组织各类文件和子目录 3. 链接文件,用"l"表示,是对文件或目录的引用 4. 设备文件,用"b"表示存储设备、用"c"表示输入输出设备

（续）

属性栏	说　明
文件权限	以三个为一组，均为rwx的三个参数的组合，其中，r代表可读（read）、w代表可写（write）、x代表可执行（execute），如果没有对应权限，则显示"-" 第一组为文件所属者具有的权限，以anaconda-ks.cfg文件为例，该文件的所属者root用户对该文件具有可读、可写、不可执行权限 第二组为文件所属组权限，即该所属组内用户具有的权限 第三组为其他用户权限，既不是文件所属者也不属于文件所属组的用户对应的权限
文件名	如果文件名以"."开头，则为隐藏文件

Linux系统中，可以使用chown、chmod指令实现不同用户对文件权限的管理，见表2-6。

表2-6　文件权限管理常用指令

指　令	指　令　介　绍	
chown	change the owner of file，变更文件或目录的所有者与所属组	
	格式	chown [选项] 用户：用户组 文件或目录 -R：同时修改目录下的所有子目录与文件的所有者与所属组
	示例	chown -R hadoop:hadoop /usr/local/hadoop #修改/usr/local/hadoop目录及其子目录下的所有文件的所有者为hadoop、所属组为hadoop
chmod	change the permissions mode of file，变更文件或目录的权限	
	格式	1. 符号形式：chmod [选项] {augo}{+-=}{rwx} 文件或目录 augo分别表示：all（所有用户）、user（所有者）、group（所属组）、other（其他用户） +-=分别表示：+（增加）、-（删除）、=（设定） 2. 数字形式：chmod [选项] xyz 文件或目录 xyz为权限的数字表示：r为4、w为2、x为1，所有者、所属组、其他用户分别将各自的三位权限值相加，分别得到x、y、z -R　同时修改目录下的所有子目录与文件的权限
	示例	chmod u=rwx,go=rx ~/.bashrc #设置~/.bashrc文件的权限为：rwxr-xr-x chmod 755 ~/.bashrc #设置~/.bashrc文件的权限为：rwxr-xr-x

（5）系统管理

系统管理常用指令详细介绍见表2-7。

扫码看视频

表2-7　系统管理常用指令

指　令	指　令　介　绍	
ps	process status，查看当前系统中正在运行的进程，配合相应参数，可以查看每个进程的进程ID（PID）、CPU占用情况、内存占用情况、进程状态等	
	格式	ps [选项] [参数]
	示例	ps -aux　#显示系统当前所有进程信息
kill	中止正在运行的进程	
	格式	kill [选项] PID -9：立即中止进程，正在进行的修改将不会被保存

（续）

指　令	指　令　介　绍	
systemctl	管理系统中的服务	
	格式	systemctl [命令选项] 服务名称 命令选项：start、restart、reload、stop、status、enable、disable 服务名称：表示管理的服务名称，例如网络服务为network
	示例	systemctl start network #启动网络服务 systemctl stop firewalld #关闭防火墙 systemctl enable sshd #设置SSH服务开机启动

任务1　安装操作系统

扫码看视频

任务描述

本任务要求王工为客户的服务器安装Linux操作系统。为了能在Windows环境中模拟安装Linux操作系统，需要使用虚拟化工具创建虚拟机，然后在虚拟机中安装Linux操作系统。

任务分析

1. 任务目标

1）了解虚拟机的概念。

2）熟练使用虚拟化工具创建虚拟机。

3）熟练安装CentOS Stream 9操作系统。

2. 任务环境

操作系统：Windows 10/Windows 11

预装软件：无

相关软件：VMware Workstation player 16.0

安装包：CentOS Stream 9操作系统镜像文件

3. 任务导图

任务导图如图2-4所示。

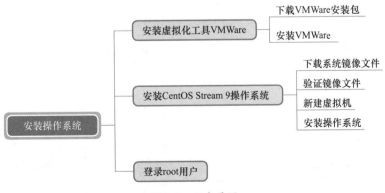

图2-4　任务导图

笔记

虚拟机（虚拟主机）：

在一台物理主机的操作系统中用虚拟化工具（如VMWare）模拟一台计算机，这台虚拟主机可以像真正的主机一样安装操作系统，用户就像在使用真正的计算机一样使用它。

笔记

VMWare系列虚拟化软件：

1）VMware Workstation：用于在 Linux 或 Windows PC 上运行虚拟机，允许用户同时创建和运行多个虚拟机。有Pro与Player版本，Pro版本功能多但是需要收费，Player版本免费，面向教育与非营利机构。

2）VMWare Fusion：面向苹果计算机的虚拟化工具，与VMWare类似。

注意

请务必使用正版软件，以尊重每一位软件开发人员的辛苦付出。本教材使用免费的VMware Workstation Player软件，如对VMware Workstation Pro版本感兴趣，可自行下载试用或付费，操作过程类似。

任务实施

1. 安装虚拟化工具VMWare

（1）下载VMWare安装包

进入VMWare官网https://www.vmware.com/cn.html，单击右上角"资源"→"产品下载"。进入所有产品下载界面，搜索"VMware Workstation"，进入VMware Workstation Player下载界面，单击"DOWNLOAD NOW"，如图2-5所示。

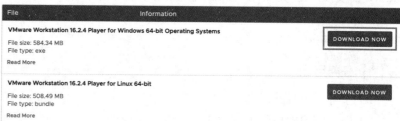

图2-5 VMware Workstation Player下载界面

（2）安装VMWare

双击安装包，进入VMware Workstation Player安装界面。按照提示，接受许可协议、选择安装位置、选择是否检测产品更新、是否加入客户体验提升计划、是否创建桌面与菜单的快捷方式，最后单击"安装"按钮等待安装，如图2-6所示。

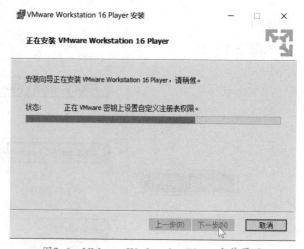

图2-6 VMware Workstation Player安装界面

安装完成后，单击"完成"按钮，桌面上会创建桌面快捷方式，如图2-7所示。

双击快捷方式图标，打开VMware Workstation Player界面，如图2-8所示。

图2-7　VMware Workstation Player快捷方式

> ⚠️ **注意**
> 由于VMware Workstation Player是免费产品，因此，不需要许可证也可以直接使用。在用户首次打开软件时，会提示"输入许可证"，可直接单击"继续"按钮，如图2-9所示。

图2-9　VMware Workstation Player首次登录提示

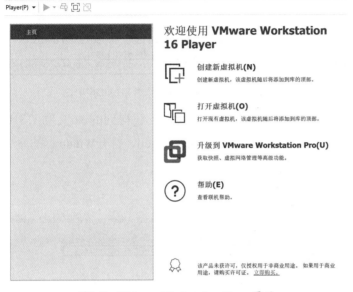

图2-8　VMware Workstation Player界面

以上，Windows平台的虚拟化工具VMWare安装完成。

2. 安装CentOS Stream 9 操作系统

（1）下载系统镜像文件

进入CentOS Stream的官网镜像下载地址：http://mirror.stream.centos.org，依次单击"9-stream/"→"BaseOS/"→"x86_64/"→"iso/"，查看文件列表，如图2-10所示。

图2-10　下载镜像文件及SHA256码

> 📝 **说明**
>
> **CentOS 版本：**
>
> CentOS 当前有两个主要版本：CentOS Linux 和 CentOS Stream（未来CentOS 项目重心）。
>
> CentOS Linux 当前代表版本为CentOS Linux 7，发布于2014年，是目前最主流及稳定的发行版，将在2024年6月30日停止支持。
>
> CentOS Stream 当前最新版本为CentOS Stream 9，发布于2021年12月3日，预计支持到2027年。
>
> 本书采用 CentOS Stream 9 作为大数据任务的基础环境。
>
> 当前，1+x 大数据运维考试环境为CentOS 7，本书对差异点做补充说明。

> ⚠ **注意**
>
> CentOS Stream 9 的镜像文件较大，经过复杂的网络传输后，可能会出现丢包等错误，需要对文件进行验证，避免出现其他复杂问题。

> ⚠ **注意**
>
> 不同 iso 文件的 SHA256 码不同，请注意务必在下载列表中下载与 iso 文件对应的 SHA256 码文件。本书中的 SHA256 码仅作参考，以官方下载链接中的 SHA256 码为准。

> ⚠ **注意**
>
> 指令中的"文件名"可以通过拖动文件到 powershell 命令行窗口的方式快速得到，且不容易出错。

下载最新的 .iso 文件，如 CentOS-Steam-9-20220928.0-x86_64-dvd1.iso 及相应的 SHA256SUM 文件。文件名中包含镜像文件发布的日期信息，以实际页面显示为准。

（2）验证镜像文件

使用记事本，打开下载得到的 SHA256SUM 文件，查看官方提供的 iso 文件的 SHA256 码，如图 2-11 所示。每个 iso 文件都有自己独立的 SHA256 码，以实际下载文件为准。

```
# CentOS-Stream-9-20220928.0-x86_64-dvd1.iso: 9003073536 bytes
SHA256 (CentOS-Stream-9-20220928.0-x86_64-dvd1.iso) =
fee69d6fcb8cd992732d71955dc37d65f451c7049327e55defd2562910b7f5c5
```

图 2-11　官方提供的 iso 文件 SHA256 码

在 Windows 中，使用指令查看下载得到的 iso 文件的 SHA256 码。按快捷键 <Win+R>，输入 powershell，打开 Windows 的命令行界面，使用如下指令，如图 2-12 所示。

```
certutil -hashfile 文件名 SHA256
```

```
PS C:\Users\admin> certutil -hashfile C:\Users\admin\Downloads\CentOS-Stream-9-202209
28.0-x86_64-dvd1.iso SHA256
SHA256 的 C:\Users\admin\Downloads\CentOS-Stream-9-20220928.0-x86_64-dvd1.iso 哈希:
fee69d6fcb8cd992732d71955dc37d65f451c7049327e55defd2562910b7f5c5
CertUtil: -hashfile 命令成功完成。
```

图 2-12　下载得到的 iso 文件 SHA256 码

将下载得到的 iso 文件的 SHA256 码与官方提供的 SHA256 码进行比对，如果一致，则说明镜像文件无误，如果不一致，则需要重新下载镜像文件。

（3）新建虚拟机

打开 VMware Workstation，单击"创建新虚拟机"，如图 2-13 所示。

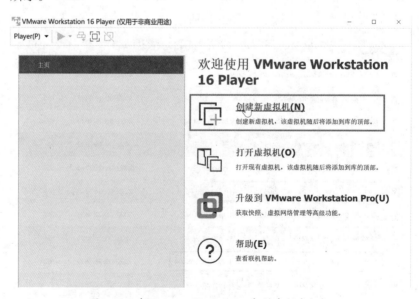

图 2-13　在 VMware Workstation 中创建新虚拟机

在"安装程序光盘映像文件"下单击"浏览"按钮,选择CentOS Stream 9镜像文件,如图2-14所示。

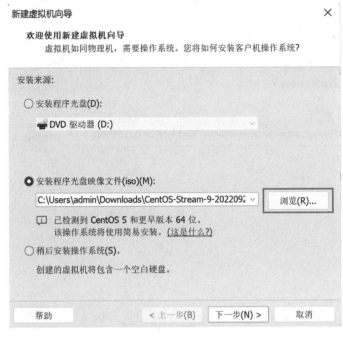

图2-14 选择镜像文件

设置用户名及密码,如图2-15所示。

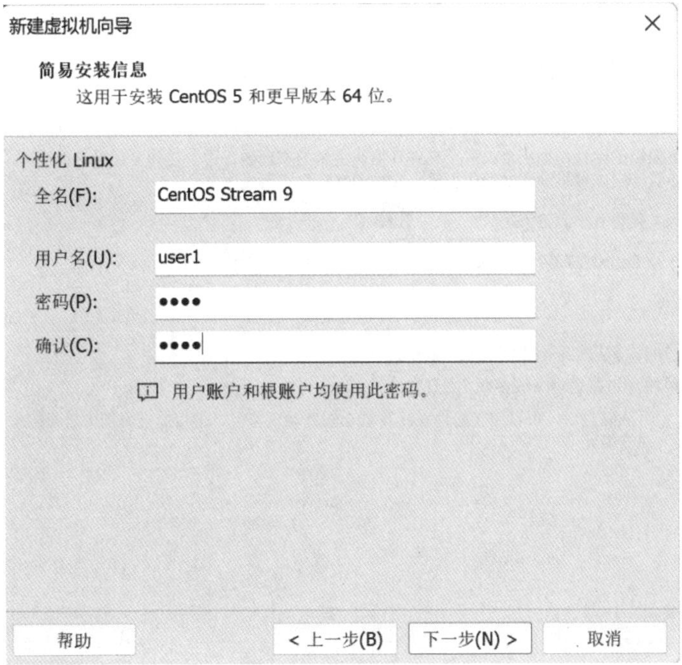

图2-15 设置用户密码

⚠ **注意**

此处的密码即为root用户密码。为了系统的安全性,请务必为root用户设置一个复杂的密码,并牢记密码。

设置虚拟机名称、存储位置,单击"下一步"按钮,如图2-16所示。

设置磁盘容量作为任务使用,可设置为20GB。可选择"将虚拟磁盘拆分成多个文件",单击"下一步"按钮,如图2-17所示。

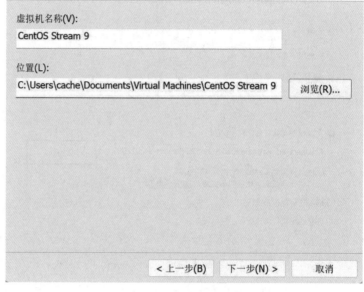

图2-16　设置虚拟机名称及存储位置

 说明

安装过程的配置参数仅供参考，可按个人使用需求设置。

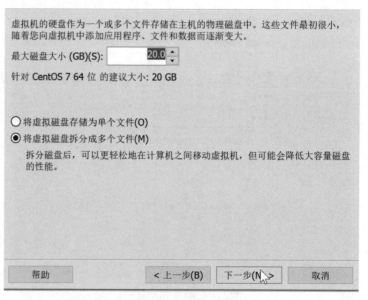

图2-17　设置虚拟机磁盘

单击"自定义硬件"按钮，如图2-18所示。

单击"内存"，设置为"2GB"。单击"处理器"，设置内核数量为"2"。设置完成后单击"关闭"按钮，如图2-19所示。

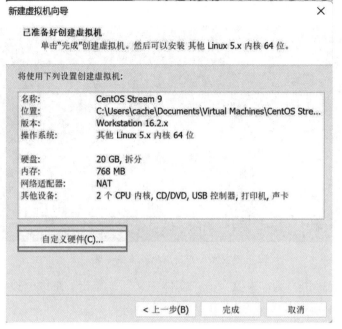

图2-18 自定义硬件

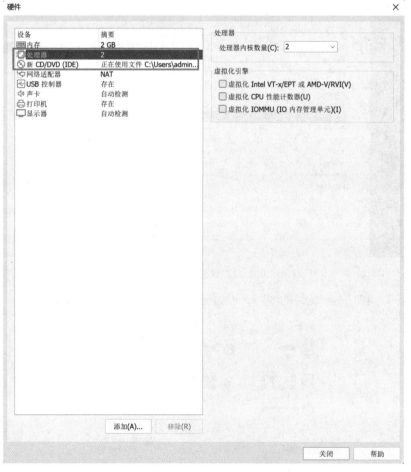

图2-19 虚拟机内存设置

（4）安装操作系统

VMware自动进入系统安装过程，如图2-20所示。

⚠ 注意

如果VMware没有进入系统安装过程，且出现图2-23所示提示，则是由于物理机未开启虚拟化功能，需要重新启动物理机，并进入BIOS界面，找到Configuration选项，选择Intel Virtual Technology，如图2-24所示，并按<Enter>键。

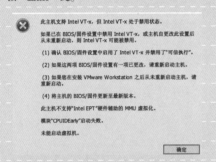

图2-23　加载镜像文件时出错

图2-24　虚拟化设置1

选择Enabled，如图2-25所示，然后按<Enter>键，保存设置后，退出BIOS。

图2-25　虚拟化设置2

图2-20　开始虚拟机系统安装

选择语言为"中文"→"简体中文"，单击"继续"按钮，如图2-21所示。

图2-21　设置虚拟机语言

单击"安装目的地"，设置磁盘分区，如图2-22所示。

图2-22　安装信息摘要之设置安装目的地

选择"自动",单击"完成"按钮,如图2-26所示。

图2-26 设置安装目标位置

单击"root密码",设置root用户密码,如图2-27所示,如果在前面的步骤中已经完成了root密码设置,则此处不需要再设置。

图2-27 安装信息摘要之设置root密码

连续两次输入相同密码后,单击"完成"按钮,如图2-28所示。

图2-28 设置root密码

⚠️ 注意

为了系统的安全性,请务必为root用户设置一个复杂的密码,并记住密码。

单击"开始安装"按钮，如图2-29所示。

图2-29 安装信息摘要界面

安装完成后，单击"重启系统"。

首次进入系统会进入配置界面，按照提示完成配置。

3. 登录root用户

以root用户登录Linux操作系统。在登录界面，单击"未列出？"。分别输入用户名、密码，按<Enter>键，如图2-30所示。

图2-30 登录系统

单击左上角"活动"按钮，再单击下方"终端"按钮，如图2-31所示，打开命令行终端。

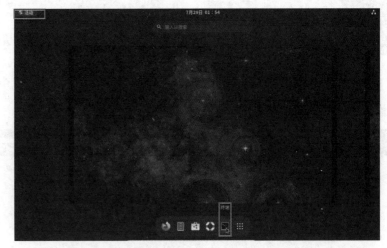

图2-31 打开命令行终端

说明

如果是命令行界面，则需要在命令行中输入用户名、密码，按<Enter>键，如图2-32所示。

```
localhost login: root
Password:
Last login: Tue Mar 23 22:54:48 on tty1
[root@localhost ~]#
```

图2-32 命令行方式登录系统

在Linux系统中，出于安全性考虑，屏幕不显示用户输入密码的过程，请注意区分密码中的大小写字母。

提示

如果不小心忘记了root密码，可以在紧急模式中修改root密码，具体方法可自行学习。

项目2 配置平台基础环境

扫码看视频

任务2 配置静态IP

任务描述

服务器通常部署于中心机房,且不配备显示设备,系统运维工程师在日常工作中都是通过IP地址远程访问服务器。为了有序管理成千上万台服务器,且每次都能通过固定IP访问到同一台服务器,服务器安装完成后的第一件事就是为服务器设置静态IP。

因此,在该任务中,王工需要为客户新安装的Linux操作系统配置静态IP。

任务分析

1. 任务目标

1)理解设置静态IP的意义。

2)按照步骤设置静态IP,并理解相关参数的意义。

2. 任务环境

操作系统:CentOS Stream 9

3. 任务导图

任务导图如图2-33所示。

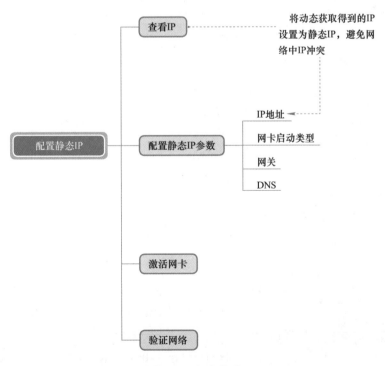

图2-33 任务导图

任务实施

1. 查看IP

使用指令查看主机IP地址。

```
# ip addr
```

观察其中ens33网卡的地址，如图2-34所示，显示ens33网卡的IP为动态（dynamic）的，地址为：192.168.245.128。

```
[root@localhost ~]# ip addr
1: lo: <LOOPBACK,UP,LOWER_UP> mtu 65536 qdisc noqueue state UNKNOWN group defaul
t qlen 1000
    link/loopback 00:00:00:00:00:00 brd 00:00:00:00:00:00
    inet 127.0.0.1/8 scope host lo
       valid_lft forever preferred_lft forever
    inet6 ::1/128 scope host
       valid_lft forever preferred_lft forever
2: ens33: <BROADCAST,MULTICAST,UP,LOWER_UP> mtu 1500 qdisc fq_codel state UP gro
up default qlen 1000
    link/ether 00:0c:29:ad:11:16 brd ff:ff:ff:ff:ff:ff
    altname enp2s1
    inet 192.168.245.128/24 brd 192.168.245.255 scope global dynamic noprefixrou
te ens33
       valid_lft 1526sec preferred_lft 1526sec
    inet6 fe80::20c:29ff:fead:1116/64 scope link noprefixroute
       valid_lft forever preferred_lft forever
```

图2-34 查看主机IP

使用指令查看ens33网卡的配置文件，如图2-35所示。

```
# cd /etc/NetworkManager/system-connections
# cat ens33.nmconnection
```

```
[root@localhost ~]# cd /etc/NetworkManager/system-connections/
[root@localhost system-connections]# cat ens33.nmconnection
[connection]
id=ens33
uuid=dd1ad377-2769-3436-968b-c9635e4ab67a
type=ethernet
autoconnect-priority=-999
interface-name=ens33
timestamp=1664746245

[ethernet]

[ipv4]
method=auto

[ipv6]
addr-gen-mode=eui64
method=auto

[proxy]
```

图2-35 查看网卡配置文件

2. 配置静态IP参数

为ens33网卡配置静态IP地址，addresses需要按照自己的主机环境设置，以上一步获取的IP地址作为此处的IP地址，可避免与网络中的其他主机产生IP冲突。例如，设置为192.168.245.168/24，指令为：

```
# nmcli con mod ens33 IPv4.addresses 192.168.245.128/24
```

为ens33网卡配置网卡启动类型，设置为"manual"，指令为：

笔记

ip addr，查看网卡信息。

注意

不同主机的网卡名称可能不同。

笔记

重点关注并配置文件中 [IPv4] 下的配置项。

method：用于设置网卡的启动类型，auto 表示自动获取 IP 地址，manual 表示手动设置静态 IP 地址。

笔记

nmcli（comand-line tool for controlling NetworkManager）用于控制NetworkManager和报告网络状态的命令行工具。

addresses：指 IP 地址。IP 地址后面的数字表示主机 IP 中网络标识的位数，例如本书主机中的 192.168.245.128/24，表示前 24 位为网络标识。

```
# nmcli con mod ens33 IPv4.method manual
```

为ens33网卡配置网关（gateway），gateway也需要按照自己的主机环境设置，需要是addresses对应网段内的网关地址，例如，设置网关为192.168.245.2：

```
# nmcli con mod ens33 IPv4.gateway 192.168.245.2
```

为ens33网卡配置DNS：

```
# nmcli con mod ens33 IPv4.dns "119.29.29.29"
```

静态IP配置流程如图2-36所示。

> **gateway**：指网关。以本书主机中的192.168.245.128/24为例，主机所在的网络内的IP范围为：192.168.245.0-192.168.245.255。默认情况下，网关地址则为：192.168.245.2。
>
> **dns**：指DNS服务器地址，用于将域名与IP地址互相转换。这里的119.29.29.29是由国内DNSPod公司提供的一个公共DNS IP地址。

```
[root@localhost system-connections]# nmcli con mod ens33 ipv4.addresses 192.168.245.128/24
[root@localhost system-connections]# nmcli con mod ens33 ipv4.method manual
[root@localhost system-connections]# nmcli con mod ens33 ipv4.gateway 192.168.245.2
[root@localhost system-connections]# nmcli con mod ens33 ipv4.dns "119.29.29.29"
[root@localhost system-connections]# cat ens33.nmconnection
[connection]
id=ens33
uuid=dd1ad377-2769-3436-968b-c9635e4ab67a
type=ethernet
autoconnect-priority=-999
interface-name=ens33
timestamp=1664808131

[ethernet]

[ipv4]
address1=192.168.245.128/24,192.168.245.2
dns=119.29.29.29;
method=manual

[ipv6]
addr-gen-mode=eui64
method=auto

[proxy]
```

> ⚠ **注意**
>
> DNS可以根据自己的需求，设置网络中可用的DNS IP，设置时不能丢失引号。

图2-36　静态IP配置流程

通过nmcli配置网卡后，网卡的配置文件内容同步更新。

3. 激活网卡

使用指令激活网卡连接：

```
# nmcli con up ens33
```

激活后，将看到ens33网卡没有提示"动态（dynamic）"，说明此时已经是静态IP了，如图2-37所示。

```
[root@localhost system-connections]# nmcli con up ens33
连接已成功激活（D-Bus 活动路径：/org/freedesktop/NetworkManager/ActiveConnection/2）
[root@localhost system-connections]# ip addr
1: lo: <LOOPBACK,UP,LOWER_UP> mtu 65536 qdisc noqueue state UNKNOWN group default qlen 1000
    link/loopback 00:00:00:00:00:00 brd 00:00:00:00:00:00
    inet 127.0.0.1/8 scope host lo
       valid_lft forever preferred_lft forever
    inet6 ::1/128 scope host
       valid_lft forever preferred_lft forever
2: ens33: <BROADCAST,MULTICAST,UP,LOWER_UP> mtu 1500 qdisc fq_codel state UP group default qlen 1000
    link/ether 00:0c:29:ad:11:16 brd ff:ff:ff:ff:ff:ff
    altname enp2s1
    inet 192.168.245.128/24 brd 192.168.245.255 scope global noprefixroute ens33
       valid_lft forever preferred_lft forever
    inet6 fe80::20c:29ff:fead:1116/64 scope link noprefixroute
       valid_lft forever preferred_lft forever
```

图2-37　网卡配置完成效果

4. 验证网络

验证静态IP的正确性与可用性，通过以下指令尝试连接网络任意地址，例如访问百度。

```
ping www.baidu.com
```

如果看到返回信息的格式如图2-38所示，则说明主机不仅是静态IP，且可以正常访问网络。

```
[root@localhost system-connections]# ping www.baidu.com
PING www.a.shifen.com (180.101.49.11) 56(84) 比特的数据。
64 比特，来自 180.101.49.11 (180.101.49.11): icmp_seq=6 ttl=128 时间=16.4 毫秒
64 比特，来自 180.101.49.11 (180.101.49.11): icmp_seq=7 ttl=128 时间=15.6 毫秒
64 比特，来自 180.101.49.11 (180.101.49.11): icmp_seq=8 ttl=128 时间=15.7 毫秒
64 比特，来自 180.101.49.11 (180.101.49.11): icmp_seq=9 ttl=128 时间=15.6 毫秒
^C
--- www.a.shifen.com ping 统计 ---
已发送 9 个包, 已接收 4 个包, 55.5556% packet loss, time 8124ms
rtt min/avg/max/mdev = 15.602/15.849/16.418/0.331 ms
```

图2-38 成功连接百度

如果无法正常连接，则依次检查IP地址、网关、DNS设置是否正确。

> ⚠️ **注意**
> 该指令不会自行停止，可按快捷键<Ctrl+C>结束指令。

任务3 远程登录

扫码看视频

任务描述

客户的服务器部署于中心机房，现已完成了静态IP配置，王工可以离开机房，在自己的Windows操作系统的计算机上通过远程连接方式，登录服务器中的Linux操作系统。

本任务先实现利用SSH服务的远程密码登录，再实现远程免密登录，提高远程访问的安全性。

任务分析

1. 任务目标

1）了解SSH定义。

2）熟练掌握远程登录。

3）熟练掌握远程文件传输。

4）熟练掌握SSH免密登录。

2. 任务环境

操作系统：Windows 10/Windows 11

相关软件：VMWare Player 16.0（内装虚拟机操作系统CentOS Stream 9）、PowerShell

3. 任务导图

任务导图如图2-39所示。

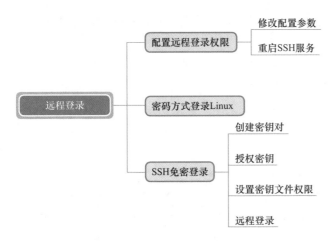

图2-39 任务导图

任务实施

1. 配置远程登录权限

在Linux系统中，使用root用户打开配置文件，并设置允许root用户远程登录。

\# vim /etc/ssh/sshd_config

输入指令，检索关键词：

/PermitRootLogin

找到图2-40所示内容。

```
#LoginGraceTime 2m
#PermitRootLogin prohibit-password
#StrictModes yes
#MaxAuthTries 6
#MaxSessions 10
```

图2-40 PermitRootLogin原配置

删除前面的#，使#后面的内容生效，并设置属性值为yes，如图2-41所示。

```
#LoginGraceTime 2m
PermitRootLogin yes
#StrictModes yes
#MaxAuthTries 6
#MaxSessions 10
```

图2-41 PermitRootLogin修改后配置

按<ESC>键，输入:wq，保存并退出配置文件。重启SSH服务。

⚠ **注意**

出于系统安全性考虑，系统默认不允许root用户远程登录，因此，如果不设置PermitRootLogin，则连接过程中将提示：Permission denied。

ⓘ **提示**

输入检索关键词指令后，按<Enter>键，再按<N>键，可以依次由上到下查看文档中符合条件的关键词。

✎ **笔记**

systemctl：管理系统中的服务。
格式：systemctl [命令选项] 服务名称
示例：
systemctl start network # 启动网络服务
systemctl stop firewalld # 关闭防火墙
systemctl enable sshd # 设置SSH服务开机启动

```
# systemctl restart sshd
```

2. 密码方式登录Linux

在Windows系统中，使用快捷键<Win+R>，输入"powershell"，打开PowerShell。

输入远程登录指令，其中IP为自己的Linux服务器IP：

```
ssh root@服务器IP
```

第一次登录时，需要输入"yes"，并按<Enter>键。

此时登录过程需要输入用户密码，输入用户密码并按<Enter>键。当出现Linux操作系统中的命令提示符，如图2-42所示，说明远程登录成功。

```
PS C:\Users\admin> ssh root@192.168.245.128
The authenticity of host '192.168.245.128 (192.168.245.128)' can't be established.
ECDSA key fingerprint is SHA256:KwdK3aeI9KfyyXkcxNuuLXim0hWC5IG04/CR5YG+eXM.
Are you sure you want to continue connecting (yes/no/[fingerprint])? yes
Warning: Permanently added '192.168.245.128' (ECDSA) to the list of known hosts.
root@192.168.245.128's password:
Activate the web console with: systemctl enable --now cockpit.socket

Last login: Sat Oct 29 21:37:58 2022 from 192.168.245.1
[root@localhost ~]#
```

图2-42　远程登录

但此时登录过程需要输入用户密码，并通过网络传输到达服务器端，显然这是不安全的。

3. SSH免密登录

通过上一步已经确定Windows能通过SSH远程登录Linux了，在此基础上，继续实现Windows向Linux免密登录。

使用指令退出Linux会话，如图2-43所示，退出登录后，命令提示符改为Windows下的提示符。

```
# exit
```

```
PS C:\Users\admin> ssh root@192.168.245.128
root@192.168.245.128's password:
Activate the web console with: systemctl enable --now cockpit.socket

Last login: Sat Oct 29 21:28:16 2022 from 192.168.245.1
[root@localhost ~]# exit
注销
Connection to 192.168.245.128 closed.
PS C:\Users\admin>
```

图2-43　退出远程登录

（1）创建密钥对

在PowerShell中，为root用户创建一个无密码的密钥对：

```
ssh-keygen -t rsa
```

密钥对生成过程中，系统提示密钥对存放位置，按<Enter>键，

📝 **笔记**

什么是SSH？

SSH 是 Secure Shell 的缩写，是专为远程登录会话和其他网络服务提供安全性的协议。利用 SSH 协议可以有效防止远程管理过程中的信息泄露问题。

📝 **笔记**

通过 SSH 协议远程连接 Linux 主机的格式为：

ssh 用户名 @ 主机名或主机IP

例如：ssh root@192.168.245.128

这条指令的意思为：以 root 用户的身份登录 IP 地址为 192.168.245.128 的主机。

该指令格式同样适用于从 Linux、UNIX 登录 Linux 服务器的场景。

📝 **笔记**

为什么需要 SSH 免密登录？

使用用户名密码的方式远程登录时，用户名密码经过网络传输，容易被第三方截取，安全性差。使用 SSH 的非对称加密方式时，网络中仅传输由公钥加密后的随机字符串，即使被第三方截取，也无法伪装成客户端登录服务器，提高安全性。

⚠️ **注意**

在 SSH 免密登录配置中，密钥对都是在客户端生成的，且其私钥永远保存在客户端中。私钥是客户端身份的象征，务必保证其隐秘性。

即可保存在系统推荐的位置。

为了提高密钥对安全性，可以为密钥对设置密码。

密钥对生成过程如图2-44所示。

密钥对生成后，可使用ls指令查看存放位置，如图2-45所示，存放位置在图2-44中查看。

其中，id_rsa为私钥，存放在客户端；id_rsa.pub为公钥，存放在打算登录的服务器端。

```
PS C:\Users\admin> ssh-keygen -t rsa
Generating public/private rsa key pair.     系统建议的密钥对存放位置
Enter file in which to save the key (C:\Users\admin/.ssh/id_rsa):
Enter passphrase (empty for no passphrase):  可为密钥对设置密码
Enter same passphrase again:
Your identification has been saved in C:\Users\admin/.ssh/id_rsa.
Your public key has been saved in C:\Users\admin/.ssh/id_rsa.pub.
The key fingerprint is:                     系统实际存放密钥对的最终位置
SHA256:z0l/sjXs2ef0/zwjqli4EU59VWLcuHjOQn5ur8v1WRQ admin@DESKTOP-RKUIE45
The key's randomart image is:
+---[RSA 3072]----+
|            .oo. |
|            .oo. |
|           ...E  |
|         . o.o  .|
|        oS.+.+  .|
|       o o+.= = .|
|        + .+ * =..|
|         =  .X.O=|
|        o ...++B=/|
+----[SHA256]-----+
PS C:\Users\admin>
```

图2-44　生成密钥对

```
PS C:\Users\admin> ls C:\Users\admin/.ssh

    目录: C:\Users\admin\.ssh


Mode                LastWriteTime     Length Name
----                -------------     ------ ----
-a----        2022/10/29     21:33       2610 id_rsa
-a----        2022/10/29     21:33        576 id_rsa.pub
```

图2-45　查看密钥对

（2）授权密钥

把公钥内容写入到被登录的服务器上。

1）在Linux中，创建~/.ssh目录，用于存放密钥文件。

　# mkdir ~/.ssh

2）在Windows中传输公钥文件到Linux系统中root用户Home路径下的.ssh文件夹中，如图2-46所示。其中，IP为Linux的IP地址。

　scp C:\Users\cache\.ssh\id_rsa.pub root@IP: ~/.ssh

笔记

scp OpenSSH secure file copy,用于在Linux中复制文件和目录。

格式：scp [选项] 源文件 目标位置

选项：-r,递归复制整个目录。

⚠ 注意

从公钥的内容中可以看到,该公钥是用来验证主机名为DESKTOP-RKUIE45上的admin用户的远程登录请求的。由于每个人Windows的主机名、用户名不同,此处内容将不同。

```
PS C:\Users\admin> scp C:\Users\admin/.ssh/id_rsa.pub root@192.168.245.128:~/.ssh
The authenticity of host '192.168.245.128 (192.168.245.128)' can't be established.
ECDSA key fingerprint is SHA256:KwdK3aeI9KfyyXkcxNuuLXim0hWC5IG04/CR5YG+eXM.
Are you sure you want to continue connecting (yes/no/[fingerprint])?
Warning: Permanently added '192.168.245.128' (ECDSA) to the list of known hosts.
root@192.168.245.128's password:
id_rsa.pub                              100%  576    15.2KB/s   00:00
PS C:\Users\admin>
```

图2-46　传输公钥文件到Linux

3）在Linux上查看公钥文件，并将其写入~/.ssh/authorized_keys文件中，如图2-47所示。

```
# cd ~/.ssh
# cat id_rsa.pub >> authorized_keys
```

```
[root@localhost ~]# cd ~/.ssh/
[root@localhost .ssh]# cat id_rsa.pub >> authorized_keys
[root@localhost .ssh]# cat authorized_keys
ssh-rsa AAAAB3NzaC1yc2EAAAADAQABAAABgQCzhgcAqRdPS1S4seKGe+ydP
OhkpHR1ZjcVZNzcM6CEBLH8aILS/R8sfp9J9e33NlUy54PWAopjtKyupk3dN2
zdb1liVyJukMNI4TrvZU2kbT3l8wzE/HR25+wPfsmtzDNSfklXKmasMR8Q20t
RVsWzKJKa4m4yYDPX890tZ0JWV6qGPe/eva7ipXmkzyFeD06HfDshQdJ+Use6
DYxjI7eQ+GbbsG/gdBJc/ve24LcQ6iNYDKTUHkHuBbIA481qhc3Ibe9Nl6cWB
lhvilaU54mvFoX/zaJlbdHwEpAMTdU+Ah9qTrxKaAsjVRve0zrrQoK/6+2BeS
Vam7IHBlEDTYBDvsyntu/JBr0xD6QkBdGd0G+p6hB89DLN89IV7HSKhXn0AkE
HHE+jtSfWIds8h342UP06fkv9/k6lzRolfYxG0ID0ieAWlRYfmOsSdE9S57Ko
/c+kqPsCp+5FlY4Aj9kso8BoLT/zgFTeQjQlCDexrN/itThfJMWztRJ0+eChN
FE= admin@DESKTOP-RKUIE45
```

图2-47　公钥写入authorized_keys

（3）设置密钥文件权限

在Linux中，为.ssh/authorized_keys文件设置权限，如图2-48所示。

```
$ chmod 0600 ~/.ssh/authorized_keys
```

```
[root@localhost .ssh]# chmod 0600 ~/.ssh/authorized_keys
[root@localhost .ssh]# ll ~/.ssh/authorized_keys
-rw-------. 1 root root 576 10月 29 21:36 /root/.ssh/authorized_keys
```

图2-48　设置authorized_keys文件权限

（4）远程登录

在Windows中，运行远程登录指令：

```
ssh root@服务器IP
```

如图2-49所示，无需输入密码便可直接登录，即实现了"免密登录"。

```
PS C:\Users\admin> ssh root@192.168.245.128
Activate the web console with: systemctl enable --now cockpit.socket

Last login: Sat Oct 29 21:29:29 2022 from 192.168.245.1
[root@localhost ~]#
```

图2-49　SSH免密登录

拓展学习

国产操作系统

经过三十余年的研发，国产操作系统已至少有15种，几种常见的国产操作系统介绍见表2-8。

扫码看视频

表2-8　常见的国产操作系统

操作系统	简　　介
深度（deepin）	deepin自主开发了深度桌面环境、自主UI库DTK、系统设置中心、音乐播放器、视频播放器、软件中心等一系列面向普通用户的应用程序，在很多领域得到了广泛应用
红旗Linux	红旗Linux是我国较成熟的Linux操作系统。红旗Linux与日本、韩国的Linux厂商共同推出了Asianux Server，拥有完善的教育系统和认证系统
银河麒麟	银河麒麟是由国防科技大学、中软公司、联想公司、浪潮集团和民族恒星公司合作研制的闭源服务器操作系统。此操作系统是863计划重大攻关科研项目，是我国拥有自主知识产权的服务器操作系统
安超OS	安超OS是一套基于服务器架构的通用型云操作系统。为企业提供高性能、高可用、高效率及易于安装维护的IT基础设施平台，加速政府和企业上云进程，为推动企业数字化转型提供完整的一站式生态解决方案
中科方德桌面操作系统	由中科方德软件有限公司推出，适配多种国产CPU，支持x86、ARM、MIPS等主流架构，支持台式计算机、笔记本计算机、一体机及嵌入式设备等形态整机、主流硬件平台和常见外设。方德桌面操作系统还预装软件中心，已上架运维近2000款优质的国产软件及开源软件

项目小结

本项目以培养大数据运维工程师掌握Linux操作系统基础理论知识为目标，介绍了Linux操作系统及其常用指令。在此基础上，针对"配置平台基础环境"的典型工作任务，根据任务的先后顺序，分为三个子任务：安装操作系统、配置静态IP、远程登录。通过完成三个子任务，练习Linux操作系统常用指令，为后续基于Linux的大数据平台部署打下基础。

实战强化

1. 自行上网查询资料，安装CentOS 7，并在CentOS 7上配置静态IP。
2. 从一台Linux上以SSH协议登录另一台Linux。
3. 执行命令ls -l 时，某行显示：-rw-r--r-- 1 chris chris 207 jul20 11:57 mydata

1）用户chris对该文件有什么权限？

2）执行命令useradd Tom后，用户Tom对该文件具有什么权限？

3）如何使任何用户都可以读写执行该文件？

4）如何把该文件所有者改为用户root？

4. 假设你是系统管理员，需要增加一个新的用户账号zheng，为新用户设置初始密码，并删除用户账号chang。

项目 3 部署 Hadoop 框架

● 项目概述

A公司根据客户的大数据业务场景需求，开发完成了一套基于Hadoop的大数据平台项目。项目进入交付验收阶段，公司委派大数据运维工程师王工前往客户的中心机房对平台进行集群部署。

本项目针对"部署Hadoop框架"的典型工作任务，由易到难，依次完成三个工作任务：部署单机模式Hadoop、部署伪分布模式Hadoop、部署全分布模式Hadoop。

在开展任务前，需要掌握必要的理论知识：什么是集群？什么是Hadoop？三种部署模式分别指的是什么？

完成任务后，进一步了解与Hadoop框架相关的知识，加深对Hadoop框架的理解。

● 学习目标

1. 建立运维工作的规范化意识。
2. 了解集群的定义。
3. 了解Hadoop的发展历史。
4. 简单描述Hadoop项目结构及其各个组件的功能。
5. 能区分Hadoop单机模式、伪分布模式、全分布模式三种模式。
6. 熟练掌握Hadoop平台的安装部署，包括单机模式、伪分布模式、全分布模式。
7. 能利用Hadoop实现简单应用，例如：词频统计。

● 思维导图

项目思维导图如图3-1所示。

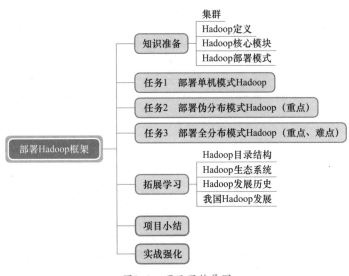

图3-1　项目思维导图

知识准备

1. 集群

集群是一组相互独立的、通过高速网络互联的计算机，它们构成了一个组，并以单一系统的模式加以管理。一个客户与集群相互作用时，集群像是一个独立的服务器。在付出较低成本的情况下获得在性能、可靠性、灵活性方面的相对较高的收益。

扫码看视频

2. Hadoop定义

Apache Hadoop项目为可靠、可扩展的分布式计算开源框架，使用简单的编程模型跨计算机集群实现对大型数据集的分布式处理。

3. Hadoop核心模块

Hadoop具有四个核心模块，见表3-1。

表3-1　Hadoop核心模块

模 块 名 称	功　　能
Hadoop Distributed File System (HDFS)	分布式文件系统，提供对应用程序数据的高吞吐量访问
Hadoop MapReduce	基于YARN的用于并行处理大数据集的系统
Hadoop YARN	是一个用于作业调度和集群资源管理的框架
Hadoop Common	是其他Hadoop模块所需的Java库和实用程序，提供文件系统和操作系统级抽象，并包含启动Hadoop所需的Java文件和脚本

4. Hadoop部署模式

Hadoop可以采用以下三种部署模式，见表3-2。

1）单机模式：默认模式，以单一Java进程运行在一台主机上，采用本地文件系统作为存储。

2）伪分布模式：每个守护进程对应单独的Java进程，运行在一台主机上。存储采用分布式文件系统HDFS，HDFS的名称节点和数据节点都在同一台机器上。

3）全分布模式：各守护进程运行在不同主机上，各主机承担集群中的不同工作。存储采用分布式文件系统HDFS。

表3-2　Hadoop部署模式

部 署 形 式	主 机 数 量	文 件 系 统
单机模式	1	本地文件系统
伪分布模式	1	分布式文件系统HDFS（单一节点）
全分布模式	N（N>1）	分布式文件系统HDFS（多节点）

任务1　部署单机模式Hadoop

扫码看视频

任务描述

王工第一次接触Hadoop，希望在Linux系统中单机部署Hadoop，并运行示例程序来学习其分布式计算模型。

任务分析

1. 任务目标

1）能理解安装Java JDK对Hadoop环境运行的意义。

2）能按照步骤部署单机模式Hadoop。

3）能利用单机模式Hadoop实现简单功能,如词频统计。

2. 任务环境

操作系统:CentOS Stream 9

软件版本:Java 1.8.0、Hadoop 3.3.4

3. 任务导图

任务导图如图3-2所示。

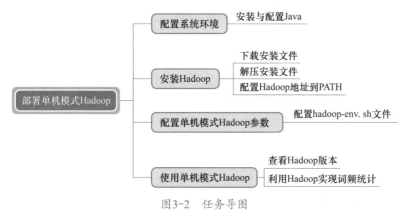

图3-2 任务导图

任务实施

1. 配置系统环境

安装与配置Java

查看系统中是否已安装Java。

```
# java -version
```

若无法查看版本信息,说明环境中未安装Java,如图3-3所示,则需安装Java。

```
[root@localhost ~]# java -version
bash: java: command not found...
Install package 'java-11-openjdk-headless' to provide command 'java'? [N/y] N
```

图3-3 java -version运行效果

系统提示,是否安装最新版本的Java 11,输入N,按<Enter>键。
若虚拟机中不是Java 8版本,则需卸载当前Java。

1）查看当前已安装的Java。

```
# rpm -qa |grep java|grep jdk
```

笔记

为什么安装Java?

Hadoop是以Java语言写成,在Hadoop的运行过程中需调用Java JDK,因此需要在服务器上预先安装Java。

说明

Hadoop的Java版本

Hadoop与Java的版本对照表见表3-3。

表3-3 Hadoop与Java的版本对照表

Hadoop版本	Java版本
3.3及更高	Java 8 & Java 11
3.0.x到3.2.x	Java 8
2.7.x到2.10.x	Java 7 & Java 8

📝 说明

虽然Hadoop 3.3及更高版本支持Java 11，但是Hadoop生态中的其他组件对Java 8的支持度更好，本书选择Java 8。

⚠️ 注意

yum指令安装过程需要保持网络状态。

2）卸载当前已安装的Java。

rpm -e --nodeps 上一步返回的包名

使用yum安装Java 1.8.0版本，如图3-4所示。

```
# yum install -y java-1.8.0-openjdk
# yum install -y java-1.8.0-openjdk-devel
```

```
[root@localhost ~]# yum install -y java-1.8.0-openjdk
上次元数据过期检查: 2:07:58 前，执行于 2023年01月03日 星期二 22时12分59秒。
依赖关系解决。
================================================================
 软件包            架构      版本              仓库         大小
================================================================
安装:
 java-1.8.0-openjdk
                  x86_64  1:1.8.0.352.b08-2.el9  appstream  459 k
安装依赖关系:
 java-1.8.0-openjdk-headless
                  x86_64  1:1.8.0.352.b08-2.el9  appstream   33 M
```
a）

```
[root@localhost ~]# yum install -y java-1.8.0-openjdk-devel
上次元数据过期检查: 2:08:52 前，执行于 2023年01月03日 星期二 22时12分59秒。
依赖关系解决。
================================================================
 软件包            架构      版本              仓库         大小
================================================================
安装:
 java-1.8.0-openjdk-devel
                  x86_64  1:1.8.0.352.b08-2.el9  appstream  9.4 M
```
b）

图3-4　安装Java

a）安装java-1.8.0-openjdk　b）安装java-1.8.0-openjdk-devel

安装完成后，查看Java版本，如图3-5所示。

```
# java -version
```

```
[root@localhost ~]# java -version
openjdk version "1.8.0_345"
OpenJDK Runtime Environment (build 1.8.0_345-b01)
OpenJDK 64-Bit Server VM (build 25.345-b01, mixed mode)
```

图3-5　查看Java版本

如果可以看到版本信息，则说明Java安装正确。

获取本机Java文件路径。

```
# which java
```

查看文件详情，若是链接文件，则查看链接文件对应的原文件，直到找到可执行文件，如图3-6所示。

```
[root@localhost local]# which java
/usr/bin/java
[root@localhost local]# ll /usr/bin/java
lrwxrwxrwx. 1 root root 22  8月 30 08:20 /usr/bin/java -> /etc/alternatives/java
[root@localhost local]# ll /etc/alternatives/java
lrwxrwxrwx. 1 root root 71  8月 30 08:20 /etc/alternatives/java -> /usr/lib/jvm/java-1.8.0-openjdk-1.8.0.345.b01-2.el9.x86_64/jre/bin/java
[root@localhost local]# ll /usr/lib/jvm/java-1.8.0-openjdk-1.8.0.345.b01-2.el9.x86_64/jre/bin/java
-rwxr-xr-x. 1 root root 18544  8月  3 11:09 /usr/lib/jvm/java-1.8.0-openjdk-1.8.0.345.b01-2.el9.x86_64/jre/bin/java
```

图3-6　查看Java文件路径

💡 建议

在实训过程中应养成"随时检查"的好习惯，确保每一步操作产生了阶段性效果。可以在一定程度上降低排错难度。

📝 笔记

文件属性中的第一个字母l表示文件为软链接类型文件。方向键指向的是源文件地址。

因此，在本书任务环境下，Java文件路径为：/usr/lib/jvm/java-1.8.0-openjdk-1.8.0.332.b09-2.e19.x86_64/jre/bin/java，对应的JAVA_HOME=/usr/lib/jvm/java-1.8.0-openjdk-1.8.0.332.b09-2.e19.x86_64/jre。

使用vim编辑器编辑/etc/profile。

```
# vim /etc/profile
```

在末行增加内容，如图3-7所示。

```
export JAVA_HOME=主机中的JAVA_HOME路径
export PATH=$PATH:$JAVA_HOME
```

```
export JAVA_HOME=/usr/lib/jvm/java-1.8.0-openjdk-1.8.0.345.b01-2.el9.x86_64/jre
export PATH=$PATH:$JAVA_HOME
```

图3-7 配置JAVA_HOME与PATH

配置完成后，按<ESC>键，输入:wq，保存并退出。

注意

使用自己主机中的Java路径。

笔记

PATH变量是一系列目录路径的集合。当用户输入指令时，操作系统便到PATH变量的路径中依次寻找该指令对应的可执行文件，然后运行该可执行文件。

设置PATH变量后，后续调用PATH路径下的可执行文件时，则不需要再输入可执行文件的完整路径信息，仅输入文件名即可。

笔记

source，重新执行脚本文件。通常用于执行刚修改过的文件，使其立即生效。

格式：source [文件]

使用source指令，使新配置生效。

```
# source /etc/profile
```

2. 安装Hadoop

（1）下载安装文件

进入Hadoop官网（https://hadoop.apache.org/）推荐的下载地址：https://dlcdn.apache.org/hadoop/common，看到Hadoop版本列表如图3-8所示。

Index of /hadoop/common

Name	Last modified	Size	Description
Parent Directory		-	
current/	2022-06-17 11:30	-	
current2/	2022-06-17 11:29	-	
hadoop-2.10.1/	2022-06-17 11:32	-	
hadoop-2.10.2/	2022-06-17 11:29	-	
hadoop-3.2.3/	2022-06-17 11:30	-	
hadoop-3.2.4/	2022-07-22 02:08	-	
hadoop-3.3.1/	2022-06-17 11:31	-	
hadoop-3.3.2/	2022-06-17 11:30	-	
hadoop-3.3.3/	2022-06-17 11:30	-	
hadoop-3.3.4/	2022-08-04 18:11	-	
stable/	2022-06-17 11:30	-	
stable2/	2022-06-17 11:29	-	
KEYS	2022-04-29 15:55	370K	
readme.txt	2015-04-21 01:32	184	

图3-8 下载列表

> **建议**
> 按照Linux系统使用的默认规范，用户安装的软件一般都是存放在"/usr/local/"目录下。

> **说明**
> 如果通过wget指令获取安装包太慢，可以使用本书提供的安装包。使用scp指令，从Windows传输文件至Linux中。

选择所需的版本，进入文件夹内，选择安装文件hadoop-*.*.*.tar.gz，例如hadoop-3.3.4.tar.gz。复制链接地址，补全下载指令，如图3-9所示。

```
# cd /usr/local
# wget https://dlcdn.apache.org/hadoop/common/hadoop-3.3.4/hadoop-3.3.4.tar.gz
```

```
[root@localhost local]# wget https://dlcdn.apache.org/hadoop/common/hadoop-3.3.4/hadoop-3.3.4.tar.gz
--2022-08-31 07:29:44--  https://dlcdn.apache.org/hadoop/common/hadoop-3.3.4/hadoop-3.3.4.tar.gz
正在解析主机 dlcdn.apache.org (dlcdn.apache.org)... 198.18.6.23
正在连接 dlcdn.apache.org (dlcdn.apache.org)|198.18.6.23|:443... 已连接。
已发出 HTTP 请求，正在等待回应... 200 OK
长度：695457782 (663M) [application/x-gzip]
正在保存至: "hadoop-3.3.4.tar.gz"

hadoop-3.3.4.tar.gz  100%[===================>] 663.24M  21.6MB/s  用时 32s

2022-08-31 07:30:17 (20.8 MB/s) - 已保存 "hadoop-3.3.4.tar.gz" [695457782/695457782])
```

图3-9 下载安装文件

查看当前目录，可以看到下载的压缩文件，如图3-10所示。

```
[root@localhost local]# ls
bin  etc  games  hadoop-3.3.4.tar.gz  include  lib  lib64  libexec  sbin  share  src
```

图3-10 查看目录

（2）解压安装文件

将安装文件解压到/usr/local目录下：

```
# tar -zxf hadoop-3.3.4.tar.gz
```

修改文件夹名称为hadoop：

```
# mv hadoop-3.3.4 hadoop
```

> **说明**
> 修改文件夹名称是为了方便后续系统管理。

完成后，可以在当前目录下查看到名为hadoop的文件夹，如图3-11所示。

```
[root@localhost local]# tar -zxf hadoop-3.3.4.tar.gz
[root@localhost local]# mv hadoop-3.3.4 hadoop
[root@localhost local]# ls
bin    games   hadoop-3.3.4.tar.gz  lib     libexec  share
etc    hadoop  include              lib64   sbin     src
```

图3-11 查看解压后的文件夹

（3）配置Hadoop地址到PATH

使用vim编辑器编辑/etc/profile。

```
# vim /etc/profile
```

在末行增加如下内容：

```
export HADOOP_HOME=/usr/local/hadoop
export PATH=$PATH:$HADOOP_HOME/bin:$HADOOP_HOME/sbin
```

配置效果如图3-12所示。

```
export JAVA_HOME=/usr/lib/jvm/java-1.8.0-openjdk-1.8.0.345.b01-5.el9.x86_64/jre
export PATH=$PATH:$JAVA_HOME

export HADOOP_HOME=/usr/local/hadoop
export PATH=$PATH:$HADOOP_HOME/bin:$HADOOP_HOME/sbin
```

图3-12 配置PATH环境变量

> **说明**
> 配置Hadoop地址到PATH后，调用Hadoop路径下的可执行文件时，无需输入可执行文件所在的路径，可提高工作效率。

配置完成后，按<ESC>键，输入:wq，保存并退出。使用source指令，使新配置生效。

```
# source /etc/profile
```

3. 配置单机模式Hadoop参数

配置hadoop-env.sh文件

使用vim编辑器编辑hadoop目录下的etc/hadoop/hadoop-env.sh文件，如图3-13所示。

```
# cd /usr/local/hadoop
# vim etc/hadoop/hadoop-env.sh
```

```
[root@localhost local]# cd /usr/local/hadoop/
[root@localhost hadoop]# vim etc/hadoop/hadoop-env.sh
```

图3-13 打开配置文件hadoop-env.sh

将JAVA_HOME前面的#去掉，并设置值为本机Java路径，如图3-14所示。

```
# Many of the options here are built from the perspective that users
# may want to provide OVERWRITING values on the command line.
# For example:
#
  JAVA_HOME=/usr/lib/jvm/java-1.8.0-openjdk-1.8.0.345.b01-2.el9.x86_64/jre
#
# Therefore, the vast majority (BUT NOT ALL!) of these defaults
# are configured for substitution and not append.  If append
# is preferable, modify this file accordingly.
```

图3-14 配置JAVA_HOME

> (i) 提示
>
> 如何快速定位JAVA_HOME？
> 在命令模式下，输入"/JAVA_HOME"并按<Enter>键。
>
> ⚠ 注意
>
> 使用自己主机中的Java路径。

设置完成后，按<ESC>键，输入:wq，保存并退出hadoop-env.sh文件。

4. 使用单机模式Hadoop

（1）查看Hadoop版本

Hadoop解压后即可使用，可以输入如下命令来检查Hadoop是否可用，成功则会显示Hadoop版本信息，如图3-15所示。

```
# hadoop version
```

```
[root@localhost hadoop]# hadoop version
Hadoop 3.3.4
Source code repository https://github.com/apache/hadoop.git -r a585a73c3e02ac62350c136643a5e7f6095a3dbb
Compiled by stevel on 2022-07-29T12:32Z
Compiled with protoc 3.7.1
From source with checksum fb9dd8918a7b8a5b430d61af858f6ec
This command was run using /usr/local/hadoop/share/hadoop/common/hadoop-common-3.3.4.jar
```

图3-15 查看Hadoop版本

> ⚠ 注意
>
> 此处，hadoop为可执行文件名称，该文件所在的文件路径为/usr/local/hadoop/bin，由于该路径已经设置在PATH中，因此此处可直接调用。若未配置PATH，则应使用指令：
>
> /usr/local/hadoop/bin/hadoop version

（2）利用Hadoop实现词频统计

利用Hadoop查找符合正则表达式的每个匹配项及匹配项个数。

说明

1）在/usr/local/hadoop路径下，创建input文件夹，作为输入目录，目录下存放的文件来自/usr/local/hadoop/etc/hadoop下的所有xml文件。

2）输出结果存放在/usr/local/hadoop/output路径下。

3）该指令使用了Hadoop自带的示例程序，对xml文件中的单词进行遍历，筛选出所有符合正则表达式dfs[a-z.]+的单词，并统计各单词出现的次数。

```
# cd /usr/local/hadoop
# mkdir input
# cp etc/hadoop/*.xml input
# hadoop jar share/hadoop/mapreduce/hadoop-mapreduce-examples-3.3.4.jar grep input output 'dfs[a-z.]+'
# cat output/*
```

运行结果如图3-16和图3-17所示。

```
[root@localhost hadoop]# mkdir input
[root@localhost hadoop]# cp etc/hadoop/*.xml input
[root@localhost hadoop]# ls input
capacity-scheduler.xml   hdfs-rbf-site.xml   kms-acls.xml       yarn-site.xml
core-site.xml            hdfs-site.xml       kms-site.xml
hadoop-policy.xml        httpfs-site.xml     mapred-site.xml
[root@localhost hadoop]# hadoop jar share/hadoop/mapreduce/hadoop-mapreduce-examples-3.3.4.jar grep input output 'dfs[a-z.]+'
2022-08-31 08:32:23,737 INFO impl.MetricsConfig: Loaded properties from hadoop-metrics2.properties
2022-08-31 08:32:23,833 INFO impl.MetricsSystemImpl: Scheduled Metric snapshot period at 10 second(s).
2022-08-31 08:32:23,833 INFO impl.MetricsSystemImpl: JobTracker metrics system started
```

图3-16　单机模式实现词频统计1

说明

在Hadoop运行过程中，可以看到屏幕中输出并行计算数据，如文件读取大小、文件写入大小、输入记录数、输出记录数等。

```
2022-08-31 08:32:26,289 INFO mapreduce.Job: Counters: 30
        File System Counters
                FILE: Number of bytes read=1202016
                FILE: Number of bytes written=3672774
                FILE: Number of read operations=0
                FILE: Number of large read operations=0
                FILE: Number of write operations=0
        Map-Reduce Framework
                Map input records=1
                Map output records=1
                Map output bytes=17
                Map output materialized bytes=25
                Input split bytes=121
                Combine input records=0
                Combine output records=0
                Reduce input groups=1
                Reduce shuffle bytes=25
                Reduce input records=1
                Reduce output records=1
                Spilled Records=2
                Shuffled Maps =1
                Failed Shuffles=0
                Merged Map outputs=1
                GC time elapsed (ms)=0
                Total committed heap usage (bytes)=728760320
        Shuffle Errors
                BAD_ID=0
                CONNECTION=0
                IO_ERROR=0
                WRONG_LENGTH=0
                WRONG_MAP=0
                WRONG_REDUCE=0
        File Input Format Counters
                Bytes Read=123
        File Output Format Counters
                Bytes Written=23
[root@localhost hadoop]# cat output/*
1       dfsadmin
```

图3-17　单机模式实现词频统计2

使用指令查看输出结果，当输出结果与图3-17一致，说明成功运行了单机模式Hadoop。

```
# cat output/*
```

说明

输出结果中，第一列为单词出现次数，第二列为符合正则表达式的单词。

项目 3 部署 Hadoop 框架

扫码看视频

任务2　部署伪分布模式Hadoop

任务描述

Hadoop最大的价值在于分布式计算，但是王工此时只有一台Linux主机，他希望能在一台主机中部署运行Hadoop集群中所需的所有守护进程（NameNode、DataNode、Secondary NameNode）。部署完成后，进一步使用Hadoop，学习其分布式计算模型。

任务分析

1. 任务目标

1）能理解使用Hadoop用户对Hadoop运行的意义。

2）能理解配置SSH免密登录对Hadoop运行的意义。

3）能简述伪分布模式下的core-site.xml、hdfs-site.xml两个配置文件中的参数意义。

4）按照步骤部署伪分布模式Hadoop。

5）能利用伪分布模式Hadoop实现简单功能，如词频统计。

2. 任务环境

操作系统：CentOS Stream 9

软件版本：Java 1.8.0、Hadoop 3.3.4

3. 任务导图

任务导图如图3-18所示。

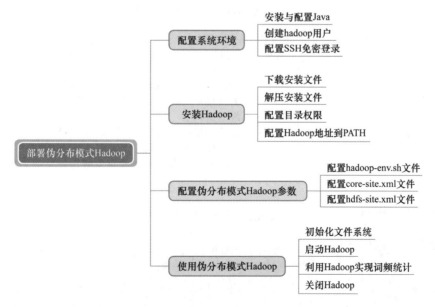

图3-18　任务导图

任务实施

1. 配置系统环境

（1）安装与配置Java

使用yum安装Java 1.8.0版本。

```
# yum install -y java-1.8.0-openjdk
# yum install -y java-1.8.0-openjdk-devel
```

安装完成后，查看Java版本：

```
# java -version
```

使用vim编辑器编辑/etc/profile，设置JAVA_HOME。

```
# vim /etc/profile
```

在末行增加如下内容：

```
export JAVA_HOME=主机中的JAVA_HOME路径
export PATH=$PATH:$JAVA_HOME
```

配置完成后，按<ESC>键，输入:wq，保存并退出。
使用source指令，使新配置生效。

```
# source /etc/profile
```

（2）创建hadoop用户

新建hadoop用户，如图3-19所示，添加完成后，可以在/etc/passwd中看到hadoop用户信息。

```
# useradd -m hadoop -s /bin/bash
```

```
[root@localhost hadoop]# useradd -m hadoop -s /bin/bash
[root@localhost hadoop]# tail -1 /etc/passwd
hadoop:x:1001:1001::/home/hadoop:/bin/bash
```

图3-19 新建hadoop用户

设置hadoop用户密码。

```
# passwd hadoop
```

根据提示输入两次密码，如图3-20所示。

```
[root@localhost hadoop]# passwd hadoop
更改用户 hadoop 的密码 。
新的密码：
无效的密码： 密码少于 8 个字符
重新输入新的密码：
passwd: 所有的身份验证令牌已经成功更新。
```

图3-20 为hadoop用户设置密码

为hadoop用户添加管理员权限，使用命令打开/etc/sudoers配置文件。

```
# visudo
```

在配置文件对应位置增加一行配置信息，如图3-21所示。

```
hadoop  ALL=(ALL)    ALL
```

说明

可在VMware Workstation中创建一台全新的Linux虚拟机开展本任务。

如果，在任务1的基础上完成本任务，注意步骤中的差异点，避免遗漏步骤。

说明

安装与配置Java过程与任务1相同，详细内容参照任务1，此处仅罗列必要步骤。

注意

使用自己主机中的Java路径。

笔记

为什么要创建hadoop用户？

为区别不同用户之间的权限、避免用户权限配置时影响其他无关用户，在环境中创建一个hadoop用户。对于一些大型软件，如MySQL，在实际企业运用中也通常为其创建单独一个用户。

建议

任务环境中可以设置简单密码方便记忆，但是在实际生产环境中，务必设置复杂密码，保证系统安全性。

笔记

为什么使用"visudo"而不使用"vi /etc/sudoers"？

使用visudo进入或退出/etc/sudoers文件时，系统会帮助检查配置文件/etc/sudoers语法，避免出错。

```
## Allow root to run any commands anywhere
root    ALL=(ALL)       ALL
hadoop  ALL=(ALL)       ALL
```

图3-21　为hadoop用户设置权限

（3）配置SSH免密登录

切换到hadoop用户：

```
# su - hadoop
```

为hadoop用户创建一个无密码的密钥对，如图3-22所示。

```
$ ssh-keygen -t rsa -P ""
```

```
[root@localhost hadoop]# su - hadoop
[hadoop@localhost ~]$ ssh-keygen -t rsa -P ""
Generating public/private rsa key pair.
Enter file in which to save the key (/home/hadoop/.ssh/id_rsa):
Created directory '/home/hadoop/.ssh'.
Your identification has been saved in /home/hadoop/.ssh/id_rsa
Your public key has been saved in /home/hadoop/.ssh/id_rsa.pub
The key fingerprint is:
SHA256:efM6IyA063+omhKSHUvUbCnunlxI1uJdSDe4ijGIKn0 hadoop@localhost.localdomain
The key's randomart image is:
+---[RSA 3072]----+
|     o o         |
|    o B o        |
|   oo = + .      |
|   = B * .       |
|  .& O + S o     |
|  B O E . o      |
|  o+ = . o       |
|  . +.. o o.     |
|   .o.o.. .o     |
+----[SHA256]-----+
```

图3-22　新建密钥对

> 配置完成后，hadoop用户只需要在指令前增加sudo关键词，即可以root身份执行指令。

📝 **笔记**

为什么需要设置 SSH 免密登录？

对于Hadoop的伪分布模式和全分布模式，Hadoop的名称节点需要登录集群中的所有节点，并启动各节点的Hadoop守护进程。为了保证安全性，Hadoop系统采用SSH协议实现远程登录。由于Hadoop并没有提供SSH输入密码登录的形式，因此，需要配置名称节点免密登录其他所有节点。

授权密钥，也就是把公钥内容写入到被登录的服务器的.ssh/authorized_keys中。在伪分布式下，则为写入本机中。执行后，可在authorized_keys文件中查看到公钥内容，如图3-23所示。

```
$ cat ~/.ssh/id_rsa.pub >> ~/.ssh/authorized_keys
$ cat ~/.ssh/authorized_keys
```

```
[hadoop@localhost ~]$ cat ~/.ssh/id_rsa.pub >>~/.ssh/authorized_keys
[hadoop@localhost ~]$ cat ~/.ssh/authorized_keys
ssh-rsa AAAAB3NzaC1yc2EAAAADAQABAAABgQCYKhmxYL0kR/nosNE9TvIUxl+xelA4lsLLISJPUoUapM5ZK
ctnHWTk8AxT7/C9mpDBugz7JJDQjvOQXyPQxUvM61nvgxXZCAknvifdfFHhXarUaWBWew4SAivonhFEYL3qKw
6CUAFAnkwbG0RNBADpELh9OldYwvRVrIK8YGFtWGjRBluXHjE4f40wbF9AtZDK/zxJS4KxPD3d5YWy6xKKrd7
4sFxsAu6LjnhjP0EouQlGhk3IUQlyi3n5/7WwmePNEyF3KLh9X6CjC2NxQzO7L29ZLCaTv0hoKs20lOEBQ3ud
ZwTDtheU69qQx8VVTdm0Ydmb/8mNM23rWp1U2zxqS2+9WKmX36MV9V6FAFsJJ0dYWKHm7AVH/ApSicnaRyrf3
8H7eEazuEjmnuHmzzjcZ2ApDlBi2wiyJlh1e18s5j0YYuoF8KnhB0kwW6C8O3K58O10O3xP1rXJRzi5Udlp+t0
qSR/NBQbx49bqhm0M5OU+WHzx3ZkbcgVtxpULewU8= hadoop@localhost.localdomain
```

图3-23　授权密钥

📝 **说明**

从密钥结尾看出，该密钥用于localhost（本地）的hadoop用户登录。

为~/.ssh/authorized_keys文件设置权限，如图3-24所示。

```
$ chmod 0600 ~/.ssh/authorized_keys
```

```
[hadoop@localhost ~]$ ll ~/.ssh/authorized_keys
-rw-r--r--. 1 hadoop hadoop 582 8月  31 09:24 /home/hadoop/.ssh/authorized_keys
[hadoop@localhost ~]$ chmod 0600 ~/.ssh/authorized_keys
[hadoop@localhost ~]$ ll ~/.ssh/authorized_keys
-rw-------. 1 hadoop hadoop 582 8月  31 09:24 /home/hadoop/.ssh/authorized_keys
```

图3-24　配置文件权限

运行指令，实现远程密钥登录本机，如图3-25所示。

```
$ ssh localhost
```

```
[hadoop@localhost ~]$ ssh localhost
The authenticity of host 'localhost (::1)' can't be established.
ED25519 key fingerprint is SHA256:8sPwPqSDVDT+nM51X5nBKW+JvYICtpuzVQAayay8bxQ.
This key is not known by any other names
Are you sure you want to continue connecting (yes/no/[fingerprint])? yes
Warning: Permanently added 'localhost' (ED25519) to the list of known hosts.
Last login: Wed Aug 31 09:22:05 2022
```

图3-25　实现密钥登录

> **说明**
> 首次登录时，需要确认连接，输入"yes"，按<Enter>键。

2. 安装Hadoop

切换回root用户。

```
$ su - root
```

（1）下载安装文件

```
# cd /usr/local
# wget https://dlcdn.apache.org/hadoop/common/hadoop-3.3.4/hadoop-3.3.4.tar.gz
# ls
```

> **说明**
> 也可以继续使用hadoop用户，只需在hadoop用户无权限执行的指令前增加sudo即可。

（2）解压安装文件

将安装文件解压到/usr/local目录下。

```
# tar -zxf hadoop-3.3.4.tar.gz
```

修改文件夹名称为hadoop。

```
# mv hadoop-3.3.4 hadoop
```

> **说明**
> （1）（2）（4）过程与任务1相同，详细内容参照任务1，此处仅罗列必要步骤。

> **说明**
> 如果通过wget指令获取安装包太慢，可以使用本书提供的安装包。使用scp指令，从Windows传输文件至Linux中。

（3）配置目录权限

修改文件权限，将文件的所有者改为hadoop用户，如图3-26所示。

```
# chown -R hadoop:hadoop ./hadoop
```

```
[root@localhost local]# ls -ld hadoop
drwxr-xr-x. 12 1024 1024 4096  8月 31 08:32 hadoop
[root@localhost local]# chown -R hadoop:hadoop ./hadoop
[root@localhost local]# ls -ld hadoop
drwxr-xr-x. 12 hadoop hadoop 4096  8月 31 08:32 hadoop
```

图3-26　配置文件夹权限

> **说明**
> 为什么需要将文件夹的所有者与所属组设置为hadoop？
> 由于在伪分布模式下，是由hadoop用户启动hadoop集群的，在启动过程中，需要进行很多读、写、执行操作，因此，需要让hadoop用户具有hadoop文件夹的最高权限。

（4）配置Hadoop地址到PATH

使用vim编辑器编辑/etc/profile。

```
# vim /etc/profile
```

在末行，增加如下内容：

```
export HADOOP_HOME=/usr/local/hadoop
export PATH=$PATH:$HADOOP_HOME/bin:$HADOOP_HOME/sbin
```

配置完成后，按<ESC>键，输入:wq，保存并退出。使用source指令，使新配置生效。

```
# source /etc/profile
```

3. 配置伪分布模式Hadoop参数

切换为hadoop用户。

```
# su - hadoop
```

> **说明**
> 接下来的内容是在hadoop路径下完成，hadoop用户具有该路径的所有权限，因此，切换成hadoop用户更好。

项目3 部署Hadoop框架

（1）配置hadoop-env.sh文件

使用vim编辑器编辑hadoop目录下的etc/hadoop/hadoop-env.sh文件。

```
$ vim /usr/local/hadoop etc/hadoop/hadoop-env.sh
```

将JAVA_HOME前面的#去掉，并设置值为本机Java路径：

```
JAVA_HOME=主机中的JAVA_HOME路径
```

设置完成后，按<ESC>键，输入:wq，保存并退出hadoop-env.sh文件。

> **说明**
> （1）过程与任务1相同，详细内容参照任务1，此处仅罗列必要步骤。

（2）配置core-site.xml文件

使用vim编辑器编辑core-site.xml文件。

```
$ vim /usr/local/hadoop/etc/hadoop/core-site.xml
```

<configuration>标签内的内容编辑成如下内容：

```xml
<configuration>
    <property>
        <name>hadoop.tmp.dir</name>
        <value>file:/usr/local/hadoop/tmp</value>
        <description>Abase for other temporary directories.</description>
    </property>
    <property>
        <name>fs.defaultFS</name>
        <value>hdfs://localhost:9000</value>
    </property>
</configuration>
```

> **笔记**
> <name>标签表示配置项名称；<value>标签标识配置项的值。

> **笔记**
> hadoop.tmp.dir用于保存临时文件，若没有配置这个参数，则默认使用的临时目录在/tmp路径下，系统重启时会被清空，影响使用。

> **笔记**
> fs.defaultFS用于指定HDFS的访问地址，端口号是官方文档中推荐的9000。

保存并退出，配置文件如图3-27所示。

```
<?xml version="1.0" encoding="UTF-8"?>
<?xml-stylesheet type="text/xsl" href="configuration.xsl"?>
<!--
  Licensed under the Apache License, Version 2.0 (the "License");
  you may not use this file except in compliance with the License.
  You may obtain a copy of the License at

    http://www.apache.org/licenses/LICENSE-2.0

  Unless required by applicable law or agreed to in writing, software
  distributed under the License is distributed on an "AS IS" BASIS,
  WITHOUT WARRANTIES OR CONDITIONS OF ANY KIND, either express or implied.
  See the License for the specific language governing permissions and
  limitations under the License. See accompanying LICENSE file.
-->

<!-- Put site-specific property overrides in this file. -->

<configuration>
    <property>
        <name>hadoop.tmp.dir</name>
        <value>file:/usr/local/hadoop/tmp</value>
        <description>Abase for other temporary directories.</description>
    </property>
    <property>
        <name>fs.defaultFS</name>
        <value>hdfs://localhost:9000</value>
    </property>
</configuration>
```

图3-27 配置core-site.xml

> **建议**
> 为了后续更好地阅读与维护配置文档，建议将标签进行缩进并对齐。

（3）配置hdfs-site.xml文件

使用vim编辑器编辑hdfs-site.xml文件。

```
$ vim /usr/local/hadoop/etc/hadoop/hdfs-site.xml
```

<configuration>标签内的内容编辑成如下内容：

```
<configuration>
    <property>
        <name>dfs.replication</name>
        <value>1</value>
    </property>
    <property>
        <name>dfs.namenode.name.dir</name>
        <value>file:/usr/local/hadoop/tmp/dfs/name</value>
    </property>
    <property>
        <name>dfs.datanode.data.dir</name>
        <value>file:/usr/local/hadoop/tmp/dfs/data</value>
    </property>
</configuration>
```

> 📝 **笔记**
> dfs.replication用于指定HDFS中同一数据的副本数量。在伪分布模式下，数据副本数量仅能设置为1。

> 📝 **笔记**
> dfs.namenode.name.dir用于指定HDFS中名称节点的文件（fsimage等）在本地文件系统中的存放位置。

> 📝 **笔记**
> dfs.namenode.data.dir用于指定HDFS中数据节点的文件在本地文件系统中的存放位置。

保存并退出，配置文件如图3-28所示。

```
<configuration>
    <property>
        <name>dfs.replication</name>
        <value>1</value>
    </property>
    <property>
        <name>dfs.namenode.name.dir</name>
        <value>file:/usr/local/hadoop/tmp/dfs/name</value>
    </property>
    <property>
        <name>dfs.datanode.data.dir</name>
        <value>file:/usr/local/hadoop/tmp/dfs/data</value>
    </property>
</configuration>
```

图3-28　配置hdfs-site.xml

> ⓘ **建议**
> 为了后续更好地阅读与维护配置文档，建议将标签进行缩进并对齐。

4．使用伪分布模式Hadoop

（1）初始化文件系统

使用如下指令，初始化文件系统。

```
$ hdfs namenode -format
```

运行结果如图3-29所示。

> 📝 **笔记**
> 伪分布模式下，Hadoop的工作需要在HDFS文件系统上完成，使用HDFS文件系统前需初始化。

> 📝 **说明**
> 指令中最前面的hdfs是/usr/local/hadoop/bin下的可执行文件的文件名，由于配置了PATH变量，所以可直接调用文件名。

```
2022-08-31 19:10:51,370 INFO common.Storage: Storage directory /usr/local/hadoop/tmp/dfs/name has been successfully formatted.
2022-08-31 19:10:51,402 INFO namenode.FSImageFormatProtobuf: Saving image file /usr/local/hadoop/tmp/dfs/name/current/fsimage.ckpt_0000000000000000000 using no compression
2022-08-31 19:10:51,505 INFO namenode.FSImageFormatProtobuf: Image file /usr/local/hadoop/tmp/dfs/name/current/fsimage.ckpt_0000000000000000000 of size 401 bytes saved in 0 seconds
2022-08-31 19:10:51,515 INFO namenode.NNStorageRetentionManager: Going to retain 1 images with txid >= 0
2022-08-31 19:10:51,549 INFO namenode.FSNamesystem: Stopping services started for active state
2022-08-31 19:10:51,550 INFO namenode.FSNamesystem: Stopping services started for standby state
2022-08-31 19:10:51,558 INFO namenode.FSImage: FSImageSaver clean checkpoint: txid=0 when meet shutdown.
2022-08-31 19:10:51,559 INFO namenode.NameNode: SHUTDOWN_MSG:
/************************************************************
SHUTDOWN_MSG: Shutting down NameNode at localhost/127.0.0.1
************************************************************/
```

图3-29　初始化文件系统

执行后将看到成功提示。若提示错误，需仔细检查各配置文件是否正确，直到成功初始化。

（2）启动Hadoop

执行指令，启动Hadoop，如图3-30所示。

```
$ start-dfs.sh
```

```
[hadoop@localhost ~]$ start-dfs.sh
Starting namenodes on [localhost]
Starting datanodes
Starting secondary namenodes [localhost.localdomain]
localhost.localdomain: Warning: Permanently added 'localhost.localdomain'
 (ED25519) to the list of known hosts.
```

图3-30　启动Hadoop

查看Java进程，如图3-31所示。

```
$ jps
```

```
[hadoop@localhost ~]$ jps
13974 NameNode
14331 SecondaryNameNode
14126 DataNode
14783 Jps
```

图3-31　启动Hadoop后执行jps

当看到NameNode、SecondaryNameNode、DataNode，说明HDFS已正常启动。

如果执行jps时，发现缺少某一个进程类型，怎么办？

1）尝试重启HDFS。

2）在HDFS上没有重要文件的前提下，可以尝试重新格式化HDFS，步骤如图3-32所示。

① 停止HDFS；

② 删除/usr/local/hadoop/tmp文件；

③ 格式化HDFS；

④ 启动HDFS。

```
$ stop-dfs.sh
$ cd /usr/local/hadoop
$ rm -rf tmp
$ hdfs namenode -format
$ start-dfs.sh
```

```
[hadoop@localhost hadoop]$ stop-dfs.sh
Stopping namenodes on [localhost]
Stopping datanodes
Stopping secondary namenodes [localhost.localdomain]
[hadoop@localhost hadoop]$ cd /usr/local/hadoop/
[hadoop@localhost hadoop]$ rm -rf tmp
[hadoop@localhost hadoop]$ hdfs namenode -format
2022-10-03 21:50:40,115 INFO namenode.NameNode: STARTUP_MSG:
/************************************************************
STARTUP_MSG: Starting NameNode
STARTUP_MSG:   host = localhost/127.0.0.1
```

图3-32　格式化文件系统

（3）利用Hadoop实现词频统计

利用Hadoop查找符合正则表达式的每个匹配项及匹配项个数。在HDFS上创建用户目录。

```
$ hdfs dfs -mkdir -p /user/hadoop
```

在HDFS上创建词频统计的输入文件夹input，并复制本地文件系统中的文件（/usr/local/hadoop/etc/hadoop/*.xml）到HDFS分布式文件系统的input文件夹内，如图3-33所示。

```
$ hdfs dfs -mkdir input
$ cd /usr/local/hadoop
$ hdfs dfs -put etc/hadoop/*.xml input
```

📝 **笔记**

jps（Java Virtual Machine Process Status Tool）：查询Linux系统当前所有Java进程pid的命令。

Hadoop由Java语言开发，运行时启动Java进程。伪分布模式下，系统将启动三个Java进程：NameNode、DataNode、Secondary NameNode。

📝 **笔记**

Hadoop正常启动后，Linux操作系统相当于有两个文件系统：本地文件系统、HDFS文件系统。

⚠️ **注意**

HDFS刚启动时，会有一段时间处于安全模式，并提示：Name node is in safe mode。该模式下，不允许用户进行任何修改文件的操作，包括上传、删除、重命名、创建等。正常情况下，只需等待一段时间，HDFS会自动结束安全模式，用户便可正常读写了。

📝 **笔记**

要使用HDFS，首先需要在HDFS中创建用户目录。例如，以hadoop用户运行时，应创建/user/hadoop目录。

```
[hadoop@localhost ~]$ hdfs dfs -mkdir -p /user/hadoop
[hadoop@localhost ~]$ hdfs dfs -mkdir input
[hadoop@localhost ~]$ hdfs dfs -ls .
Found 1 items
drwxr-xr-x   - hadoop supergroup          0 2022-08-31 19:26 input
[hadoop@localhost ~]$ hdfs dfs -put /usr/local/hadoop/etc/hadoop/*.xml input
[hadoop@localhost ~]$ hdfs dfs -ls input
Found 10 items
-rw-r--r--   1 hadoop supergroup       9213 2022-08-31 19:26 input/capacity-schedule
r.xml
-rw-r--r--   1 hadoop supergroup       1073 2022-08-31 19:26 input/core-site.xml
-rw-r--r--   1 hadoop supergroup      11765 2022-08-31 19:26 input/hadoop-policy.xml
-rw-r--r--   1 hadoop supergroup        683 2022-08-31 19:26 input/hdfs-rbf-site.xml
-rw-r--r--   1 hadoop supergroup       1118 2022-08-31 19:26 input/hdfs-site.xml
-rw-r--r--   1 hadoop supergroup        620 2022-08-31 19:26 input/httpfs-site.xml
-rw-r--r--   1 hadoop supergroup       3518 2022-08-31 19:26 input/kms-acls.xml
-rw-r--r--   1 hadoop supergroup        682 2022-08-31 19:26 input/kms-site.xml
-rw-r--r--   1 hadoop supergroup        758 2022-08-31 19:26 input/mapred-site.xml
-rw-r--r--   1 hadoop supergroup        690 2022-08-31 19:26 input/yarn-site.xml
```

图3-33　上传任务文件至HDFS

可以看到分布式文件系统中，hadoop用户目录下input文件夹里的文件为刚才复制得到的文件。

运行示例文件，实现对input目录下文件的词频统计，如图3-34所示，能看到Hadoop进行词频统计的运行数据。

```
$ cd /usr/local/hadoop
$ hadoop jar share/hadoop/mapreduce/hadoop-mapreduce-examples-3.3.4.jar grep input output 'dfs[a-z.]+'
```

> ⚠ 注意
>
> 如果要再次运行该词频统计指令，需要先删除HDFS文件系统中的output路径，否则将看到提示："hdfs://localhost:9000/user/hadoop/output already exists"。
>
> 删除output路径的指令：
>
> $ hdfs dfs -rm -r output

```
2022-08-31 19:39:43,340 INFO mapreduce.Job:  map 100% reduce 100%
2022-08-31 19:39:43,340 INFO mapreduce.Job: Job job_local8656673_
0002 completed successfully
2022-08-31 19:39:43,342 INFO mapreduce.Job: Counters: 36
        File System Counters
                FILE: Number of bytes read=1144558
                FILE: Number of bytes written=3662461
                FILE: Number of read operations=0
                FILE: Number of large read operations=0
                FILE: Number of write operations=0
                HDFS: Number of bytes read=60678
                HDFS: Number of bytes written=515
                HDFS: Number of read operations=73
                HDFS: Number of large read operations=0
                HDFS: Number of write operations=14
                HDFS: Number of bytes read erasure-coded=0
        Map-Reduce Framework
                Map input records=4
                Map output records=4
                Map output bytes=101
                Map output materialized bytes=115
                Input split bytes=132
                Combine input records=0
                Combine output records=0
                Reduce input groups=1
                Reduce shuffle bytes=115
                Reduce input records=4
                Reduce output records=4
                Spilled Records=8
                Shuffled Maps =1
                Failed Shuffles=0
                Merged Map outputs=1
                GC time elapsed (ms)=0
                Total committed heap usage (bytes)=1857028096
        Shuffle Errors
                BAD_ID=0
                CONNECTION=0
                IO_ERROR=0
                WRONG_LENGTH=0
                WRONG_MAP=0
                WRONG_REDUCE=0
        File Input Format Counters
                Bytes Read=219
        File Output Format Counters
                Bytes Written=77
```

图3-34　词频统计运行结果

在分布式文件系统上查看输出结果，如图3-35所示。

```
$ hdfs dfs -cat output/*
```

```
[hadoop@localhost ~]$ hdfs dfs -cat output/*
1       dfsadmin
1       dfs.replication
1       dfs.namenode.name.dir
1       dfs.datanode.data.dir
```

图3-35　查看运行结果

当输出结果与图3-35一致，说明成功运行了伪分布模式Hadoop，达成任务目标。

（4）关闭Hadoop

执行以下指令，关闭Hadoop，如图3-36所示。

```
$ stop-dfs.sh
```

```
[hadoop@localhost ~]$ stop-dfs.sh
Stopping namenodes on [localhost]
Stopping datanodes
Stopping secondary namenodes [localhost.localdomain]
[hadoop@localhost ~]$ jps
16214 Jps
```

图3-36　关闭Hadoop

系统依次关闭namenodes、datanodes、secondary namenodes。再次使用jps指令，将看到仅剩下jps进程。

任务3　部署全分布模式Hadoop

扫码看视频

任务描述

在实际生产环境中，为了体现分布式计算价值，Hadoop部署形式采用全分布模式，充分调动各主机上计算、内存等资源。王工经过任务1、任务2已经充分掌握了Hadoop部署的整体流程，接下来，王工正式前往客户指定的中心机房进行Hadoop全分布模式部署。

本任务中，假设有三台主机，其中一台为master、两台为slave，三台主机通过网络相连。master节点主要承载平台中HDFS的NameNode与SecondaryNameNode以及YARN资源管理器相关的ResourceManager与NodeManager；slave节点承载DataNode及NodeManager。部署全分布模式网络拓扑结构如图3-37所示。

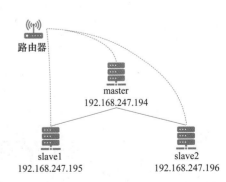

图3-37　部署全分布模式网络拓扑结构

任务分析

1. 任务目标

1）能理解配置SSH免密登录对Hadoop运行的意义。

2）能理解配置主机名对系统运维的意义。

3）能简述全分布模式下的core-site.xml、hdfs-site.xml、mapred-site.xml、yarn-site.xml配置文件中的参数意义。

4）按照步骤部署全分布模式Hadoop。

5）能简述全分布模式与伪分布模式配置流程的区别。

6）能利用全分布模式Hadoop实现简单功能，如词频统计。

2. 任务环境

主机数量：3台

操作系统：CentOS Stream 9

软件版本：Java 1.8.0、Hadoop 3.3.4

3. 任务导图

任务导图如图3-38所示。

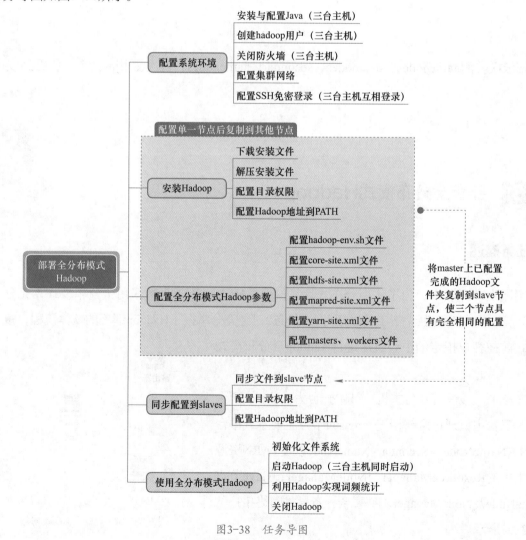

图3-38 任务导图

任务实施

1. 配置系统环境

（1）安装与配置Java（三台主机）

使用yum安装Java 1.8.0版本。

```
# yum install -y java-1.8.0-openjdk
# yum install -y java-1.8.0-openjdk-devel
```

设置JAVA_HOME。

```
# vim /etc/profile
```

在末行增加如下内容：

```
export JAVA_HOME=本机的java路径
export PATH=$PATH:$JAVA_HOME
```

使用source指令，使新配置生效。

```
# source /etc/profile
```

（2）创建hadoop用户（三台主机）

创建hadoop用户。

```
# useradd hadoop
```

设置hadoop用户密码。

```
# passwd hadoop
```

为hadoop用户添加管理员权限。

```
# visudo
```

在配置文件中增加一行配置信息。

```
hadoop  ALL=(ALL)   ALL
```

（3）关闭防火墙（三台主机）

运行指令，关闭防火墙，并设置开机不自动启动防火墙，查看防火墙状态，确认防火墙已关闭，如图3-39所示。

```
# systemctl stop firewalld
# systemctl disable firewalld
# systemctl status firewalld
```

```
[root@localhost ~]# systemctl stop firewalld
[root@localhost ~]# systemctl disable firewalld
Removed /etc/systemd/system/multi-user.target.wants/firewalld.service.
Removed /etc/systemd/system/dbus-org.fedoraproject.FirewallD1.service.
[root@localhost ~]# systemctl status firewalld
● firewalld.service - firewalld - dynamic firewall daemon
   Loaded: loaded (/usr/lib/systemd/system/firewalld.service; disabled; vendor preset: enabled)
   Active: inactive (dead)
     Docs: man:firewalld(1)

9月 02 01:04:24 localhost systemd[1]: Starting firewalld - dynamic firewall daemon...
9月 02 01:04:24 localhost systemd[1]: Started firewalld - dynamic firewall daemon.
10月 31 21:51:22 localhost.localdomain systemd[1]: Stopping firewalld - dynamic firewall daemo...
10月 31 21:51:22 localhost.localdomain systemd[1]: firewalld.service: Deactivated successfully.
10月 31 21:51:22 localhost.localdomain systemd[1]: Stopped firewalld - dynamic firewall daemon.
10月 31 21:51:22 localhost.localdomain systemd[1]: firewalld.service: Consumed 1.305s CPU time.
lines 1-11/11 (END)
```

图3-39 关闭防火墙

ⓘ 建议

本任务建议在VMware Workstation中创建三台全新的Linux虚拟机，不建议在任务2的伪分布模式基础上继续任务。

📝 说明

（1）（2）过程与任务1相同，详细内容参照任务1，此处仅罗列必要步骤。

⚠ 注意

此步骤需要在三台主机中做相同配置。

⚠ 注意

此步骤需要在三台主机中做相同配置。

📝 笔记

为什么关闭防火墙？

在完全分布式环境下，DataNode节点需要访问NameNode节点的9000端口，汇报节点状态等。关闭防火墙，可以保证各节点的端口能正常访问。

（4）配置集群网络

1）修改各机器主机名。

master主机：

```
# hostnamectl set-hostname master
```

slave1主机：

```
# hostnamectl set-hostname slave1
```

slave2主机：

```
# hostnamectl set-hostname slave2
```

master主机配置效果如图3-40所示，slave主机类似。

```
[root@localhost ~]# hostnamectl set-hostname master
[root@localhost ~]# su - root
[root@master ~]#
```

图3-40 master主机配置主机名

2）获取各主机IP地址，并将IP设置为静态IP。

```
# ip addr
#nmcli con mod ens33 ipv4.addresses 192.168.247.194/24
#nmcli con mod ens33 ipv4.method manual
#nmcli con mod ens33 ipv4.gateway 192.168.247.2
#nmcli con mod ens33 ipv4.dns "119.29.29.29"
#nmcli con up ens33
```

master主机配置效果如图3-41所示。

```
[root@master ~]# ip addr
1: lo: <LOOPBACK,UP,LOWER_UP> mtu 65536 qdisc noqueue state UNKNOWN group default qlen 1000
    link/loopback 00:00:00:00:00:00 brd 00:00:00:00:00:00
    inet 127.0.0.1/8 scope host lo
       valid_lft forever preferred_lft forever
    inet6 ::1/128 scope host
       valid_lft forever preferred_lft forever
2: ens33: <BROADCAST,MULTICAST,UP,LOWER_UP> mtu 1500 qdisc fq_codel state UP group default qlen 1000
    link/ether 00:0c:29:4e:70:d8 brd ff:ff:ff:ff:ff:ff
    altname enp2s1
    inet 192.168.247.194/24 brd 192.168.247.255 scope global dynamic noprefixroute ens33
       valid_lft 1463sec preferred_lft 1463sec
    inet6 fe80::20c:29ff:fe4e:70d8/64 scope link noprefixroute
       valid_lft forever preferred_lft forever
[root@master ~]# nmcli con mod ens33 ipv4.addresses 192.168.247.194/24
[root@master ~]# nmcli con mod ens33 ipv4.method manual
[root@master ~]# nmcli con up ens33
连接已成功激活（D-Bus 活动路径：/org/freedesktop/NetworkManager/ActiveConnection/2）
```

图3-41 master主机配置静态IP

3）依次检测三台主机是否互联。

master主机尝试连接slave1与slave2。

```
# ping 192.168.247.195
# ping 192.168.247.196
```

连接效果如图3-42所示，说明两台主机网络通信正常。

⚠ 注意

配置主机名后，需要重新登录，才能在命令提示符中看到更新后的主机名。

📝 说明

即使不配置成静态IP也是可以配置全分布模式Hadoop的，但是为了避免在主机重启后IP变更，此处对主机进行静态IP配置。

本书中，IP配置如下：

master：192.168.247.194

slave1：192.168.247.195

slave2：192.168.247.196

```
[root@master ~]# ping 192.168.247.195
PING 192.168.247.195 (192.168.247.195) 56(84) 比特的数据。
64 比特，来自 192.168.247.195: icmp_seq=1 ttl=64 时间=0.657 毫秒
64 比特，来自 192.168.247.195: icmp_seq=2 ttl=64 时间=0.321 毫秒
^C
--- 192.168.247.195 ping 统计 ---
已发送 2 个包，已接收 2 个包，0% packet loss, time 1025ms
rtt min/avg/max/mdev = 0.321/0.489/0.657/0.168 ms
[root@master ~]# ping 192.168.247.196
PING 192.168.247.196 (192.168.247.196) 56(84) 比特的数据。
64 比特，来自 192.168.247.196: icmp_seq=1 ttl=64 时间=0.610 毫秒
64 比特，来自 192.168.247.196: icmp_seq=2 ttl=64 时间=0.640 毫秒
^C
--- 192.168.247.196 ping 统计 ---
已发送 2 个包，已接收 2 个包，0% packet loss, time 1014ms
rtt min/avg/max/mdev = 0.610/0.625/0.640/0.015 ms
```

图3-42　master主机连接slave1与slave2

同样的方式，使slave1与slave2主机依次与另两台主机尝试连接。

4）设置三台主机的配置文件/etc/hosts。

\# vi /etc/hosts

在文件末尾添加以下配置，如图3-43所示。

```
192.168.247.194 master
192.168.247.195 slave1
192.168.247.196 slave2
```

```
127.0.0.1   localhost localhost.localdomain localhost4 localhost4.localdomain4
::1         localhost localhost.localdomain localhost6 localhost6.localdomain6
192.168.247.194 master
192.168.247.195 slave1
192.168.247.196 slave2
```

图3-43　/etc/hosts文件配置

（5）配置SSH免密登录（三台主机互相登录）

实现hadoop用户从master免密登录到master、slave1、slave2。

在master节点上，切换到hadoop用户。

\# su - hadoop

为hadoop用户创建一个无密码的密钥对。

$ ssh-keygen -t rsa -P ""

1）免密登录master节点。

授权密钥到master节点，将公钥写入本机的~/.ssh/authorized_keys文件中。

$ cat ~/.ssh/id_rsa.pub >> ~/.ssh/authorized_keys

为.ssh/authorized_keys文件设置权限。

$ chmod 0600 ~/.ssh/authorized_keys

运行指令，验证远程密钥登录本机。

$ ssh localhost

笔记

/etc/hosts 配置文件是做什么的？

/etc/hosts文件用于将IP与对应主机名进行映射。设置完成后，当前主机将根据主机名，自动匹配对应IP。

/etc/hosts配置完成后，可以使用"ping 主机名"的方式验证其配置的有效性，运行效果同"ping IP"。

通过这样的方式，当集群中节点的IP发生改变时，可以快速更新配置，利于集群系统的长期运维。

说明

（5）过程与任务2类似，目标是实现master免密登录到slave1与slave2。

笔记

为什么需要设置 SSH 免密登录？

对于Hadoop的伪分布模式和全分布模式，Hadoop的名称节点需要登录集群中的所有节点，并启动各节点的Hadoop守护进程。为了保证安全性，Hadoop系统采用SSH协议实现远程登录。由于Hadoop并没有提供SSH输入密码登录的形式，因此，需要配置名称节点免密登录其他所有节点。

注意

在操作过程中，注意区分指令是在哪个节点上执行的。

2）免密登录slave1与slave2节点。

分别在slave1与slave2节点的hadoop Home目录下创建.ssh目录，用于存放来自master节点的id_rsa.pub文件。

```
$ mkdir ~/.ssh
```

将master节点的公钥id_rsa.pub复制到slave1与slave2节点，在master节点运行如下指令，如图3-44所示。

```
$ scp ~/.ssh/id_rsa.pub hadoop@slave1:~/.ssh
$ scp ~/.ssh/id_rsa.pub hadoop@slave2:~/.ssh
```

```
[hadoop@master ~]$ scp ~/.ssh/id_rsa.pub hadoop@slave1:~/.ssh
The authenticity of host 'slave1 (192.168.247.200)' can't be established.
ED25519 key fingerprint is SHA256:NPLGfAgraqbnuchouFX8jSPAzFG7tZprz6J60zkHKBE.
This key is not known by any other names
Are you sure you want to continue connecting (yes/no/[fingerprint])? yes
Warning: Permanently added 'slave1' (ED25519) to the list of known hosts.
hadoop@slave1's password:
id_rsa.pub                               100%  567   814.4KB/s   00:00
[hadoop@master ~]$ scp ~/.ssh/id_rsa.pub hadoop@slave2:~/.ssh
The authenticity of host 'slave2 (192.168.247.199)' can't be established.
ED25519 key fingerprint is SHA256:NPLGfAgraqbnuchouFX8jSPAzFG7tZprz6J60zkHKBE.
This host key is known by the following other names/addresses:
    ~/.ssh/known_hosts:1: slave1
Are you sure you want to continue connecting (yes/no/[fingerprint])? yes
Warning: Permanently added 'slave2' (ED25519) to the list of known hosts.
hadoop@slave2's password:
id_rsa.pub                               100%  567   711.5KB/s   00:00
```

图3-44 master节点复制公钥至slave1、slave2

复制成功后，在slave1与slave2节点的~/.ssh路径下存有来自master的hadoop用户密钥对的公钥文件id_rsa.pub。将id_rsa.pub内容复制到authorized_keys文件中。

slave1与slave2节点分别运行指令。

```
$ cat ~/id_rsa.pub >>~/.ssh/authorized_keys
$ cat ~/.ssh/authorized_keys
$ chmod 0600 ~/.ssh/authorized_keys
```

运行结果如图3-45所示。

```
[hadoop@slave1 ~]$ cat ~/.ssh/authorized_keys
ssh-rsa AAAAB3NzaC1yc2EAAAADAQABAAABgQCsC0M5NdwW5gVo4BNmJebMySg
6ysWkTt6BEUHOlLdVLyHEZxxzaW8vDDnXOFnz6KLZlICMJcP2hlL7H9FI4armSD
8zVeEIAwnLzXOntrrXQfh0LHnBVoG/3Voiu8yL7FgtPPVAAoBS42u/TwOnuxYe6
OyL21cJl3luxp9QCh4QdwAsyRLM/ItUrvJ7oWbRW0lMCI/le5m7i3rG4j6FHdy/
DyL+tzoNu832XFNVwv76xza5QlJWRPGY0mlar8IIwu7HZPMb+clNXSLi3HCzpZa
HtyzKaSPM7sUX5mzf2tVcShwpCoPtyWtnVjfC72N66swxD8kuw+LctVVaaGYmxa
/XKbS3djj9WlH5Ncyq0GEM/s64mgdEkjqtdCByooKRP34AdtC2b9KJcjqSv5JYD
g0wBGf49/JhjiwNzeBvoEkDVoGOtiDABKkDtayZYSMJm+Lddx6BJclpmn1YOPDC
JYlQvM+ZKW1+6b2eYkvwZhS0GgQgGzu0o7q+aDuAhvYHk0= hadoop@master
```

a)

```
[hadoop@slave2 ~]$ cat ~/.ssh/authorized_keys
ssh-rsa AAAAB3NzaC1yc2EAAAADAQABAAABgQCsC0M5NdwW5gVo4BNmJebMySg6y
sWkTt6BEUHOlLdVLyHEZxxzaW8vDDnXOFnz6KLZlICMJcP2hlL7H9FI4armSD8zVe
EIAwnLzXOntrrXQfh0LHnBVoG/3Voiu8yL7FgtPPVAAoBS42u/TwOnuxYe6OyL21c
Jl3luxp9QCh4QdwAsyRLM/ItUrvJ7oWbRW0lMCI/le5m7i3rG4j6FHdy/DyL+tzoN
u832XFNVwv76xza5QlJWRPGY0mlar8IIwu7HZPMb+clNXSLi3HCzpZaHtyzKaSPM7
sUX5mzf2tVcShwpCoPtyWtnVjfC72N66swxD8kuw+LctVVaaGYmxa/XKbS3djj9Wl
H5Ncyq0GEM/s64mgdEkjqtdCByooKRP34AdtC2b9KJcjqSv5JYDg0wBGf49/Jhjiw
NzeBvoEkDVoGOtiDABKkDtayZYSMJm+Lddx6BJclpmn1YOPDCJYlQvM+ZKW1+6b2e
YkvwZhS0GgQgGzu0o7q+aDuAhvYHk0= hadoop@master
```

b)

图3-45 slave节点的authorized_keys
a) slave1节点 b) slave2节点

📝 说明

slave1与slave2节点的authorized_keys中都包含了hadoop@master，说明slave1与slave2节点接受来自master上的hadoop用户的免密登录。

验证master到slave1与slave2节点免密登录，如图3-46所示。

```
[hadoop@master ~]$ ssh slave1
[hadoop@slave1 ~]$ exit
[hadoop@master ~]$ ssh slave2
[hadoop@slave2 ~]$ exit
```

```
[hadoop@master ~]$ ssh slave1
Last login: Thu Sep  1 02:33:40 2022 from ::1
[hadoop@slave1 ~]$ exit
注销
Connection to slave1 closed.
[hadoop@master ~]$ ssh slave2
Last login: Thu Sep  1 02:33:53 2022
[hadoop@slave2 ~]$ exit
注销
Connection to slave2 closed.
```

图3-46　master节点分别登录slave1、slave2

2. 安装Hadoop

在master节点，切换回root用户，完成以下配置：

```
$ su - root
```

（1）下载安装文件

```
# cd /usr/local
# wget https://dlcdn.apache.org/hadoop/common/hadoop-3.3.4/hadoop-3.3.4.tar.gz
```

（2）解压安装文件

将安装文件解压到/usr/local目录下。

```
# tar -zxf hadoop-3.3.4.tar.gz
```

修改文件夹名称为hadoop：

```
# mv hadoop-3.3.4 hadoop
```

（3）配置目录权限

修改文件权限，将文件的所有者改为hadoop用户。

```
# chown -R hadoop:hadoop ./hadoop
```

（4）配置Hadoop地址到PATH

使用vim编辑器编辑/etc/profile。

```
# vim /etc/profile
```

在末行增加如下内容：

```
export HADOOP_HOME=/usr/local/hadoop
export PATH=$PATH: $HADOOP_HOME/bin:$HADOOP_HOME/sbin
```

使用source指令，使新配置生效。

```
# source /etc/profile
```

> **注意**
> master登录slave1后需要先退出登录，再尝试登录slave2。否则，直接运行ssh slave2则是由slave1连接slave2。

> **说明**
> ssh slave1指令中，没有指定远程登录slave1的用户名，则表示使用当前系统用户名登录slave1。

> **说明**
> 也可以继续使用hadoop用户，只需在hadoop用户无权限的指令前增加sudo即可。

> **说明**
> （1）（2）（3）（4）过程与任务2相同，详细内容参照任务2，此处仅罗列必要步骤。

> **说明**
> 如果通过wget指令获取安装包太慢，可以使用本书提供的安装包。使用scp指令，从Windows传输文件至Linux中。

> **说明**
> 为什么需要将文件夹的所有者与所属组设置为hadoop？
> 由于在全分布模式下，是由hadoop用户启动Hadoop集群的，在启动过程中，需要进行很多读、写、执行操作，因此，需要让hadoop用户具有hadoop文件夹的最高权限。

扫码看视频

3. 配置全分布模式Hadoop参数

在master节点，切换为hadoop用户，完成（1）～（6）配置。

```
# su - hadoop
```

（1）配置hadoop-env.sh文件

使用vim编辑器编辑hadoop目录下的etc/hadoop/hadoop-env.sh文件。

```
$ vim /usr/local/hadoop/etc/hadoop/hadoop-env.sh
```

将JAVA_HOME前面的#去掉，并设置值为本机Java路径：

```
JAVA_HOME=本机JAVA_HOME路径
```

设置完成后，按<ESC>键，输入:wq，保存并退出hadoop-env.sh文件。

（2）配置core-site.xml文件

使用vim编辑器编辑core-site.xml文件。

```
$ vim /usr/local/hadoop/etc/hadoop/core-site.xml
```

<configuration>标签内的内容编辑成如下内容：

```
<configuration>
    <property>
        <name>Hadoop.tmp.dir</name>
        <value>file:/usr/local/Hadoop/tmp</value>
        <descrIPtion>Abase for other temporary directories.
</descrIPtion>
    </property>
    <property>
        <name>fs.defaultFS</name>
        <value>hdfs://master:9000</value>
    </property>
    <property>
        <name>io.file.buffer.size</name>
        <value>131072</value>
    </property>
</configuration>
```

保存并退出，配置文件如图3-47所示。

```
<configuration>
    <property>
        <name>hadoop.tmp.dir</name>
        <value>file:/usr/local/hadoop/tmp</value>
        <description>Abase for other temporary directories.</description>
    </property>
    <property>
        <name>fs.defaultFS</name>
        <value>hdfs://master:9000</value>
    </property>
    <property>
        <name>io.file.buffer.size</name>
        <value>131072</value>
    </property>
</configuration>
```

图3-47　配置core-site.xml

说明

接下来的内容是在hadoop路径下设置，hadoop用户具有该路径权限，且hadoop用户需具备当前路径下新建、修改的所有文件权限。因此，切换成hadoop用户更好。

说明

（1）过程与任务1相同，详细内容参照任务1，此处仅罗列必要步骤。

笔记

core-site.xml文件用于定义Hadoop系统级别的参数。

笔记

hadoop.tmp.dir用于保存临时文件，若没有配置这个参数，则默认使用的临时目录在/tmp路径下，系统重启时会被清空，影响使用（同伪分布）。

笔记

fs.defaultFS用于指定HDFS的访问地址，端口号是官方文档中推荐的9000。

笔记

io.file.buffer.size用于指定流文件的缓冲区大小，默认值4096。

（3）配置hdfs-site.xml文件

使用vim编辑器编辑hdfs-site.xml文件。

```
$ vim /usr/local/hadoop/etc/hadoop/hdfs-site.xml
```

<configuration>标签内的内容编辑成如下内容：

```xml
<configuration>
    <property>
        <name>dfs.replication</name>
        <value>3</value>
    </property>
    <property>
        <name>dfs.namenode.name.dir</name>
        <value>file:/usr/local/hadoop/tmp/dfs/name</value>
    </property>
    <property>
        <name>dfs.datanode.data.dir</name>
        <value>file:/usr/local/hadoop/tmp/dfs/data</value>
    </property>
</configuration>
```

> 📝 **笔记**
> hdfs-site.xml文件用于设置HDFS参数，包括文件存放位置、副本数量等。

> 📝 **笔记**
> dfs.replication用于指定HDFS中同一数据的副本数量。在全分布模式下，数据副本数量默认为3。

> 📝 **笔记**
> dfs.namenode.name.dir用于指定HDFS中名称节点的文件（fsimage等）在本地文件系统中的存放位置（同伪分布）。

> 📝 **笔记**
> dfs.datanode.data.dir用于指定HDFS中数据节点的文件在本地文件系统中的存放位置（同伪分布）。

保存并退出，配置文件如图3-48所示。

```xml
<configuration>
    <property>
        <name>dfs.replication</name>
        <value>3</value>
    </property>
    <property>
        <name>dfs.namenode.name.dir</name>
        <value>file:/usr/local/hadoop/tmp/dfs/name</value>
    </property>
    <property>
        <name>dfs.datanode.data.dir</name>
        <value>file:/usr/local/hadoop/tmp/dfs/data</value>
    </property>
</configuration>
```

图3-48　配置hdfs-site.xml

（4）配置mapred-site.xml文件

编辑mapred-site.xml文件。

```
$ vim /usr/local/hadoop/etc/hadoop/mapred-site.xml
```

<configuration>标签内的内容编辑成如下内容：

```xml
<configuration>
    <property>
        <name>mapreduce.framework.name</name>
        <value>yarn</value>
    </property>
    <property>
        <name>mapreduce.jobhistory.address</name>
        <value>master:10020</value>
    </property>
    <property>
        <name>mapreduce.jobhistory.webapp.address</name>
        <value>master:19888</value>
    </property>
</configuration>
```

> 📝 **笔记**
> mapred-site.xml文件用于设置MapReduce参数，包括作业历史服务器、应用程序参数等。

> 📝 **笔记**
> mapreduce.framework.name用于指定集群的资源管理框架，指定为yarn，则使用YARN集群来实现资源的分配。

> 📝 **笔记**
> mapreduce.jobhistory.address定义历史服务器的地址和端口，通过历史服务器查看已经运行完的Mapreduce作业记录。

> 📝 **笔记**
> mapreduce.jobhistory.webapp.address定义历史服务器Web应用访问的地址和端口。

保存并退出，配置文件如图3-49所示。

```xml
<configuration>
    <property>
        <name>mapreduce.framework.name</name>
        <value>yarn</value>
    </property>
    <property>
        <name>mapreduce.jobhistory.address</name>
        <value>master:10020</value>
    </property>
    <property>
        <name>mapreduce.jobhistory.webapp.address</name>
        <value> master:19888</value>
    </property>
</configuration>
```

图3-49 配置mapred-site.xml

（5）配置yarn-site.xml文件

使用vim编辑器编辑yarn-site.xml文件。

```
$ vim /usr/local/hadoop/etc/hadoop/yarn-site.xml
```

将<configuration>标签内的内容编辑成如下内容：

```xml
<configuration>
    <property>
        <name>yarn.resourcemanager.address</name>
        <value>master:8032</value>
    </property>
    <property>
        <name>yarn.resourcemanager.scheduler.address</name>
        <value>master:8030</value>
    </property>
    <property>
        <name>yarn.resourcemanager.resource-tracker.address</name>
        <value>master:8031</value>
    </property>
    <property>
        <name>yarn.resourcemanager.admin.address</name>
        <value>master:8033</value>
    </property>
    <property>
        <name>yarn.resourcemanager.webapp.address</name>
        <value>master:8088</value>
    </property>
    <property>
        <name>yarn.nodemanager.aux-services</name>
        <value>mapreduce_shuffle</value>
    </property>
    <property>
        <name>yarn.nodemanager.aux-services.mapreduce.shuffle.class</name>
        <value>org.apache.hadoop.mapred.ShuffleHandler</value>
    </property>
    <property>
        <name>yarn.application.classpath</name>
```

📝 笔记

yarn-site.xml文件用于设置集群资源管理系统参数，包括ResourceManager和NodeManager通信端口、Web监控端口等。

📝 笔记

1）yarn.resourcemanager.address提供给客户端访问的地址。客户端通过该地址向RM提交应用程序，杀死应用程序等。

2）yarn.resourcemanager.scheduler.address是ResourceManager对ApplicationMaster暴露的访问地址，ApplicationMaster通过这个地址向RM申请资源，释放资源。

3）yarn.resourcemanager.resource-tracker.address指定ResourceManager 提供给NodeManager的地址。NodeManager通过该地址向ResourceManager汇报"心跳"，领取任务等。

4）yarn.resourcemanager.admin.address指定ResourceManager 提供给管理员的访问地址。管理员通过该地址向RM发送管理命令等。

5）yarn.resourcemanager.webapp.address指定ResourceManager对Web服务提供地址。用户可通过该地址在浏览器中查看集群各类信息。

6）yarn.nodemanager.aux-services，通过该配置项，用户可以自定义一些服务，此处定义里Map-Reduce的shuffle功能。

7）yarn.application.classpath，配置hadoop依赖项的路径。

```xml
        <value>/usr/local/hadoop/etc/hadoop:/usr/local/hadoop/share/hadoop/common/lib/*:/usr/local/hadoop/share/hadoop/common/*:/usr/local/hadoop/share/hadoop/hdfs:/usr/local/hadoop/share/hadoop/hdfs/lib/*:/usr/local/hadoop/share/hadoop/hdfs/*:/usr/local/hadoop/share/hadoop/mapreduce/*:/usr/local/hadoop/share/hadoop/yarn:/usr/local/hadoop/share/hadoop/yarn/lib/*:/usr/local/hadoop/share/hadoop/yarn/*</value>
    </property>
</configuration>
```

> **注意**
> 这里配置的value为以下命令的执行结果：
> $ hadoop classpath
> 请配置为自己主机的返回值。

保存并退出，配置文件如图3-50所示。

```xml
<configuration>
    <property>
        <name>yarn.resourcemanager.address</name>
        <value>master:8032</value>
    </property>
    <property>
        <name>yarn.resourcemanager.scheduler.address</name>
        <value>master:8030</value>
    </property>
    <property>
        <name>yarn.resourcemanager.resource-tracker.address</name>
        <value>master:8031</value>
    </property>
    <property>
        <name>yarn.resourcemanager.admin.address</name>
        <value>master:8033</value>
    </property>
    <property>
        <name>yarn.resourcemanager.webapp.address</name>
        <value>master:8088</value>
    </property>
    <property>
        <name>yarn.nodemanager.aux-services</name>
        <value>mapreduce_shuffle</value>
    </property>
    <property>
        <name>yarn.nodemanager.aux-services.mapreduce.shuffle.class</name>
        <value>org.apache.hadoop.mapred.ShuffleHandler</value>
    </property>
    <property>
        <name>yarn.application.classpath</name>
        <value>/usr/local/hadoop/etc/hadoop:/usr/local/hadoop/share/hadoop/common/lib/*:/usr/local/hadoop/share/hadoop/common/*:/usr/local/hadoop/share/hadoop/hdfs:/usr/local/hadoop/share/hadoop/hdfs/lib/*:/usr/local/hadoop/share/hadoop/hdfs/*:/usr/local/hadoop/share/hadoop/mapreduce/*:/usr/local/hadoop/share/hadoop/yarn:/usr/local/hadoop/share/hadoop/yarn/lib/*:/usr/local/hadoop/share/hadoop/yarn/*</value>
    </property>
</configuration>
```

图3-50　配置yarn-site.xml

（6）配置masters、workers文件

新建masters文件。

```
# cd /usr/local/hadoop/etc/hadoop
# vi masters
```

配置内容如下：

```
master
```

编辑workers文件。

```
# vi workers
```

> **说明**
> Hadoop 3.0版本前，slave列表由slaves文件管理，3.0版本起，salves文件改名为workers，文件配置方式相同。

扫码看视频

配置内容如下：

```
master
slave1
slave2
```

保存并退出，如图3-51所示。

```
[hadoop@master hadoop]$ cd etc/hadoop/
[hadoop@master hadoop]$ vi masters
[hadoop@master hadoop]$ vi workers
[hadoop@master hadoop]$ cat masters
master
[hadoop@master hadoop]$ cat workers
master
slave1
slave2
```

图3-51 配置masters、workers

4. 同步配置到slaves

（1）同步文件到slave节点

将master上的Hadoop安装文件同步到slave1、slave2，如图3-52所示。

```
$ scp -r /usr/local/hadoop/ root@slave1:/usr/local/
$ scp -r /usr/local/hadoop/ root@slave2:/usr/local/
```

```
[hadoop@master hadoop]$ scp -r /usr/local/hadoop/ root@slave2:/usr/local
root@slave2's password:
```

图3-52 同步Hadoop至slave2

在slave1、slave2的/usr/local路径下可查看到新增的hadoop文件夹，内容与master上一致。

（2）配置目录权限

分别在两个slave节点切换到root用户，分别设置hadoop文件夹权限，设置其所属者与所属组为hadoop用户。

```
$ su - root
# cd /usr/local
# chown -R hadoop:hadoop ./hadoop
```

（3）配置Hadoop地址到PATH

分别在两个slave节点编辑/etc/profile。

```
# vim /etc/profile
```

增加如下内容：

```
export HADOOP_HOME=/usr/local/hadoop
export PATH=$PATH:$HADOOP_HOME/bin:$HADOOP_HOME/sbin
```

使用source指令，使新配置生效。

```
# source /etc/profile
```

5. 使用全分布模式Hadoop

（1）初始化文件系统

在master节点，初始化文件系统。

```
$ hdfs namenode -format
```

> **说明**
>
> 默认情况下，slave1、slave2主机未开通root用户远程登录权限。进入/etc/ssh/sshd_config文件，删除PermitRootLogin前的#，配置属性值为yes，如图3-53所示。
>
> ```
> #LoginGraceTime 2m
> PermitRootLogin yes
> #StrictModes yes
> ```
>
> 图3-53 配置PermitRootLogin
>
> 重启sshd服务：systemctl restart sshd

> **说明**
>
> 初始化成功后，master节点上/usr/local/hadoop路径下出现tmp文件夹。

（2）启动Hadoop（三台主机同时启动）

在master节点启动Hadoop集群。启动完成后，用jps查看各节点Java进程，如图3-54所示。

```
$ start-all.sh
$ jps
```

```
[hadoop@master hadoop]$ start-all.sh
WARNING: Attempting to start all Apache Hadoop daemons as hadoop in 10 seconds.
WARNING: This is not a recommended production deployment configuration.
WARNING: Use CTRL-C to abort.
Starting namenodes on [master]
Starting datanodes
Starting secondary namenodes [master]
Starting resourcemanager
Starting nodemanagers
[hadoop@master hadoop]$ jps
48257 SecondaryNameNode
48772 Jps
48583 NodeManager
47929 NameNode
48060 DataNode
48478 ResourceManager
```
a）

```
[hadoop@slave1 hadoop]$ jps      [hadoop@slave2 hadoop]$ jps
38101 DataNode                   37762 Jps
38220 NodeManager                37620 NodeManager
38366 Jps                        37502 DataNode
```
b）   c）

图3-54　查看Hadoop集群状态
a）master节点　b）slave1节点　c）slave2节点

> **笔记**
> 只需要在master节点执行start-all.sh，slave节点会同时启动。第一次启动后，slave节点上/usr/local/hadoop路径下会出现tmp文件夹。

> **笔记**
> Hadoop正常启动后，Linux操作系统相当于有两个文件系统：本地文件系统、HDFS文件系统。

（3）利用Hadoop实现词频统计

利用Hadoop查找符合正则表达式的每个匹配项及匹配项个数。

在HDFS上创建词频统计的输入文件夹input，并复制本地文件系统中的文件（/usr/local/hadoop/etc/hadoop/*.xml）到HDFS分布式文件系统的input文件夹内。

```
$ cd /usr/local/hadoop
$ hdfs dfs -mkdir -p /user/hadoop/input
$ hdfs dfs -put etc/hadoop/*.xml input
```

运行示例文件，实现对HDFS文件系统内input目录下的所有文件的词频统计。

```
$ hadoop jar share/hadoop/mapreduce/hadoop-mapreduce-examples-3.3.4.jar grep input output 'dfs[a-z.]+'
```

在分布式文件系统上查看输出结果，如图3-55所示。

```
$ hdfs dfs -cat output/*
```

```
[hadoop@master hadoop]$ hdfs dfs -cat output/*
1    dfsadmin
1    dfs.replication
1    dfs.namenode.name.dir
1    dfs.datanode.data.dir
```

图3-55　查看运行结果

当输出结果与图3-55一致，说明成功运行了全分布模式Hadoop，达成任务目标。

> **注意**
> 文件上传后，可观察各节点中tmp路径下的文件。

> **注意**
> 如果要再次运行该词频统计指令，需要先删除HDFS文件系统中的output路径，否则将看到提示："hdfs://master:9000/user/hadoop/output already exists"。删除output路径：
>
> $ hdfs dfs -rm -r output

（4）关闭Hadoop

执行以下指令，关闭Hadoop，如图3-56所示。

```
$ stop-all.sh
```

```
[hadoop@master hadoop]$ stop-all.sh
WARNING: Stopping all Apache Hadoop daemons as hadoop in 10 seconds.
WARNING: Use CTRL-C to abort.
Stopping namenodes on [master]
Stopping datanodes
Stopping secondary namenodes [master]
Stopping nodemanagers
slave1: WARNING: nodemanager did not stop gracefully after 5 seconds: Trying to kill with kill -9
slave2: WARNING: nodemanager did not stop gracefully after 5 seconds: Trying to kill with kill -9
Stopping resourcemanager
[hadoop@master hadoop]$ jps
10026 Jps
```

图3-56　关闭Hadoop

在所有节点中，再次使用jps指令，将看到仅剩下jps进程。

拓展学习

1. Hadoop目录结构

Hadoop安装目录包括 bin、etc、include、lib、libexec、sbin、share共7个目录以及一些其他文件，如图3-57所示。下面简单介绍下各目录内容及作用。

扫码看视频

```
[root@master hadoop]# ll
总用量 104
drwxr-xr-x. 2 hadoop hadoop  4096 7月  29 06:44 bin
drwxr-xr-x. 3 hadoop hadoop    20 7月  29 05:35 etc
drwxr-xr-x. 2 hadoop hadoop   106 7月  29 06:44 include
drwxr-xr-x. 3 hadoop hadoop    20 7月  29 06:44 lib
drwxr-xr-x. 4 hadoop hadoop  4096 7月  29 06:44 libexec
-rw-rw-r--. 1 hadoop hadoop 24707 7月  28 13:30 LICENSE-binary
drwxr-xr-x. 2 hadoop hadoop  4096 7月  29 06:44 licenses-binary
-rw-rw-r--. 1 hadoop hadoop 15217 7月  16 11:20 LICENSE.txt
drwxr-xr-x. 3 hadoop hadoop  4096 10月 22 08:46 logs
-rw-rw-r--. 1 hadoop hadoop 29473 7月  16 11:20 NOTICE-binary
-rw-rw-r--. 1 hadoop hadoop  1541 4月  22  2022 NOTICE.txt
-rw-rw-r--. 1 hadoop hadoop   175 4月  22  2022 README.txt
drwxr-xr-x. 3 hadoop hadoop  4096 7月  29 05:35 sbin
drwxr-xr-x. 4 hadoop hadoop    31 7月  29 07:21 share
```

图3-57　Hadoop目录结构

1）bin：存放操作Hadoop相关服务（HDFS、YARN）的脚本。

2）etc：存放Hadoop配置文件，主要包含core-site.xml、hdfs-site.xml、mapred-site.xml、yarn-site.xml等配置文件。

3）include：对外提供的编程库头文件，这些头文件均是用C++定义的，通常用于C++程序访问HDFS或者编写MapReduce程序。

4）lib：包含了Hadoop对外提供的编程动态库和静态库，与include目录中的头文件结合使用。

5）libexec：各个服务对应的Shell配置文件所在的目录，可用于配置日志输出、启动参数等基本信息。

6）sbin：存放Hadoop管理脚本，主要包含HDFS和YARN中各类服务的启动/关闭脚本。

7）share：Hadoop各个模块编译后的jar包。

2. Hadoop生态系统

Hadoop构建了一套丰富的生态系统，如图3-58所示，部分组件简介见表3-4。

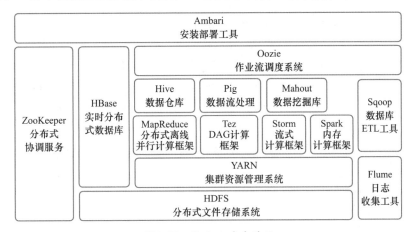

图3-58　Hadoop生态系统

表3-4　Hadoop生态系统部分组件介绍

组　件	介　绍
Ambari	用于配置、管理和监控Apache Hadoop集群，包括对Hadoop HDFS、Hadoop MapReduce、Hive、HBase、ZooKeeper、Oozie、Pig和Sqoop的支持。Ambari还提供了一个仪表板，用于查看集群健康状况，并能够以用户友好的方式诊断其性能
ZooKeeper	分布式应用程序的高性能协调服务
HBase	一个可扩展的分布式数据库，支持大型表的结构化和非结构化数据存储
Oozie	一个可扩展、高可靠的工作流调度系统，用于管理Apache Hadoop作业，工作流作业是一个有向无环图（DAG）。支持MapReduce、Pig、Hive、Sqoop等
Hive	提供数据汇总和即席查询的数据仓库基础设施
Pig	用于并行计算的高级数据流语言和执行框架
Mahout	可扩展的机器学习和数据挖掘库，目的是让数学家、数据科学家、统计学家快速实现目标算法
Tez	基于Hadoop YARN构建的通用数据流编程框架，它提供了一个强大而灵活的引擎来执行任意DAG任务，实现批处理和交互式用例。Tez应用于Hadoop生态系统中的Hive、Pig和其他框架以及其他商业软件（例如ETL工具），以取代Hadoop MapReduce作为底层执行引擎
Storm	分布式实时计算系统，可以实时、轻松、可靠地处理数据流，用于实时分析、在线机器学习、连续计算、分布式RPC、ETL等。具有速度快、可扩展、高容错的特点
Spark	用于Hadoop数据的快速通用内存计算引擎。Spark支持广泛的应用程序，包括ETL、机器学习、流处理和图形计算
Sqoop	用于在Hadoop与传统数据库（mysql、postgresql等）进行数据的传递，可以将一个关系型数据库中的数据导入到Hadoop的HDFS中，也可以将HDFS的数据导入到关系型数据库中
Flume	一个高可用、高可靠分布式海量日志采集、聚合和传输系统，支持在日志系统中定制各类数据发送方，用于收集数据；同时提供对数据进行简单处理，并写到各种数据接受方的能力

3. Hadoop发展历史

Hadoop已发展20多年，其发展历史如图3-59所示。

图3-59　Hadoop发展历史

4. 我国Hadoop发展

在Hadoop的开发过程中，我国华为公司为此做了巨大贡献。2011年，也就是Hadoop发布的那一年，Hadoop公司公布了Hadoop项目贡献者数据分析，可以看到各公司的Hadoop参与情况，如图3-60所示。其中，华为公司在Hadoop重要贡献公司名单内，并且排在Google和Cisco的前面，说明华为公司一直在积极参与开源社区贡献。

Hadoop在国内的应用主要以互联网公司为主，包括百度、阿里巴巴、腾讯等。

2019年7月25日，阿里巴巴发布了我国自主研发的计算引擎——飞天大数据平台，是当时全球集群规模最大的计算平台，最大可扩展至10万台计算集群，支持海量数据存储和计算，如图3-61所示。该平台实现Serverless大数据服务模式，将大数据能力作为一种付费资源提供给企业，减少企业本地服务器部署压力，提升企业工作效率、减少开发和人力资源的成本投入、专注于业务发展，培养更多业务型技术人员。

《超越Hadoop》书中提到：阿里云战略上最与众不同之处，就是坚持追求拥有自己的具有竞争力的核心技术。在越来越关注我国自主研发核心技术的今天，飞天大数据平台具有强大的时代意义。在我国数字化发展过程中，飞天大数据平台成为其背后坚实的技术支持，实现让老百姓办事"最多跑一次"、给城市红绿灯"装上大脑"等，促进数字化转型，提升广大群众的幸福感。

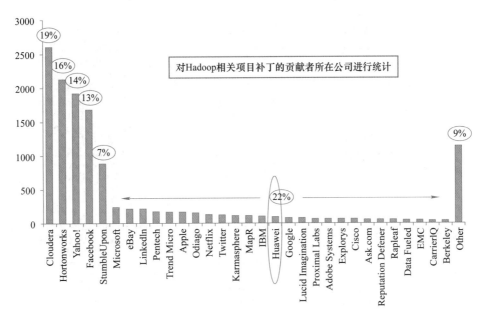

图3-60　Hadoop重要贡献公司名单

图3-61　飞天大数据产品家族

项目小结

本项目围绕"部署Hadoop框架"典型工作任务，以大数据运维工程师新人视角，讲解了集群、Hadoop、Hadoop核心模块、Hadoop部署形式等基本概念，并由浅入深逐步展开三个任务，层层递进地学习如何部署Hadoop框架，三个任务横向对比如图3-62所示，引导大数据运维工程师新人逐步掌握部署Hadoop框架的相关能力。在完成任务后，开展拓展知识学习，包括Hadoop目录结构、生态系统、发展历史，最后引入我国与Hadoop发展的相关内容。

	任务1：部署单机模式	任务2：部署伪分布模式	任务3：部署全分布模式
配置系统环境	安装配置Java	安装配置Java 创建hadoop用户 配置SSH免密登录	安装配置Java（三台主机） 创建hadoop用户（三台主机） 关闭防火墙（三台主机） 配置集群网络 配置SSH互相免密登录
安装Hadoop	下载、解压 配置PATH	下载、解压 配置目录权限 配置PATH	下载、解压 配置目录权限 ⎫ master节点 配置PATH
配置参数	配置hadoop-env.sh	配置hadoop-env.sh 配置core-site.xml 配置hdfs-site.xml	配置hadoop-env.sh 配置core-site.xml 配置hdfs-site.xml 配置mapred-site.xml ⎫ master节点 配置yarn-site.xml 配置masters、workers
同步配置			同步Hadoop至slave节点 配置目录权限 ⎫ slave节点 配置PATH
使用Hadoop （词频统计）	执行词频统计 （本地文件系统）	初始化HDFS 启动Hadoop 执行词频统计 （HDFS文件系统） 关闭Hadoop	初始化HDFS 启动Hadoop（三台主机同时启动） 执行词频统计 （HDFS文件系统） 关闭Hadoop

图3-62 项目任务横向对比总结

实战强化

1. 运维人员在/etc/profile文件中设置了JAVA_HOME变量，使用source /etc/profile指令使变量生效，运行结果如图3-63所示。

```
[hadoop@master ~]$ sudo vim /etc/profile
[sudo] hadoop 的密码：
[hadoop@master ~]$ source /etc/profile
bash: export: "=": 不是有效的标识符
bash: export: "/usr/lib/jvm/java-1.8.0-openjdk-1.8.0.345.b01-2.el9.x86_64
/jre": 不是有效的标识符
```

图3-63 source /etc/profile报错图

已知其/etc/profile文件的部分内容，如图3-64所示。

```
export JAVA_HOME = /usr/lib/jvm/java-1.8.0-openjdk-1.8.0.345.b01-2.el9.x8
6_64/jre
export HADOOP_HOME=/usr/local/hadoop
export PATH=$PATH:$JAVA_HOME:$HADOOP_HOME/bin:$HADOOP_HOME/sbin
```

图3-64 /etc/profile内容截图

试分析系统报错的原因，并给出解决方案。

2. 运维人员在进行Hadoop伪分布模式部署时，需要配置Hadoop的core-site.xml文件，输入以下指令，如图3-65所示。

```
[hadoop@localhost ~]$ cd /usr/local/hadoop
[hadoop@localhost hadoop]$ vim core-site.xml
```

图3-65　错误指令图

系统显示该文件内容为空，如图3-66所示。

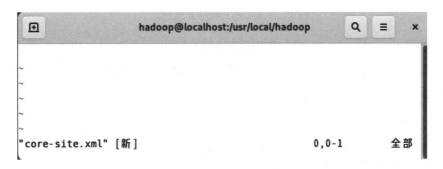

图3-66　错误响应图

但是其他运维人员进入core-site.xml文件时能看到内容，如图3-67所示。

```
<?xml version="1.0" encoding="UTF-8"?>
<?xml-stylesheet type="text/xsl" href="configuration.xsl"?>
<!--
  Licensed under the Apache License, Version 2.0 (the "License");
  you may not use this file except in compliance with the License.
  You may obtain a copy of the License at

    http://www.apache.org/licenses/LICENSE-2.0

  Unless required by applicable law or agreed to in writing, software
  distributed under the License is distributed on an "AS IS" BASIS,
  WITHOUT WARRANTIES OR CONDITIONS OF ANY KIND, either express or implied.
  See the License for the specific language governing permissions and
  limitations under the License. See accompanying LICENSE file.
-->

<!-- Put site-specific property overrides in this file. -->

<configuration>
</configuration>
```

图3-67　正确响应图

试分析该运维人员出错的原因，并给出解决方案。

3. 运维人员在部署伪分布模式Hadoop时，对HDFS进行格式化时出现错误，如图3-68所示。

```
Caused by: com.ctc.wstx.exc.WstxParsingException: Unexpected close tag </property>; expected </
name>.
 at [row,col,system-id]: [31,14,"file:/usr/local/hadoop/etc/hadoop/hdfs-site.xml"]
        at com.ctc.wstx.sr.StreamScanner.constructWfcException(StreamScanner.java:634)
        at com.ctc.wstx.sr.StreamScanner.throwParseError(StreamScanner.java:504)
        at com.ctc.wstx.sr.StreamScanner.throwParseError(StreamScanner.java:488)
        at com.ctc.wstx.sr.BasicStreamReader.reportWrongEndElem(BasicStreamReader.java:3352)
        at com.ctc.wstx.sr.BasicStreamReader.readEndElem(BasicStreamReader.java:3279)
        at com.ctc.wstx.sr.BasicStreamReader.nextFromTree(BasicStreamReader.java:2900)
        at com.ctc.wstx.sr.BasicStreamReader.next(BasicStreamReader.java:1121)
        at org.apache.hadoop.conf.Configuration$Parser.parseNext(Configuration.java:3396)
        at org.apache.hadoop.conf.Configuration$Parser.parse(Configuration.java:3182)
        at org.apache.hadoop.conf.Configuration.loadResource(Configuration.java:3075)
        ... 10 more
[hadoop@localhost hadoop]$
```

图3-68　格式化HDFS报错

通过阅读错误提示，该运维人员应该如何排查错误？

4. 运维人员部署全分布模式Hadoop集群，使用start-all.sh启动Hadoop集群时，收到系统如图3-69所示的反馈信息。

```
[hadoop@master hadoop]$ start-all.sh
WARNING: Attempting to start all Apache Hadoop daemons as hadoop in 10 seconds.
WARNING: This is not a recommended production deployment configuration.
WARNING: Use CTRL-C to abort.
Starting namenodes on [master]
Starting datanodes
slave1: hadoop@slave1: Permission denied (publickey,gssapi-keyex,gssapi-with-mic,password).
slave2: hadoop@slave2: Permission denied (publickey,gssapi-keyex,gssapi-with-mic,password).
Starting secondary namenodes [master]
Starting resourcemanager
Starting nodemanagers
slave1: hadoop@slave1: Permission denied (publickey,gssapi-keyex,gssapi-with-mic,password).
slave2: hadoop@slave2: Permission denied (publickey,gssapi-keyex,gssapi-with-mic,password).
```

图3-69　启动Hadoop集群

试分析该反馈信息的意思，并分析其原因，最终给出解决方案。

项目 4 使用 HDFS

项目概述

A公司根据客户的大数据业务场景需求，开发完成了一套基于Hadoop的大数据平台项目。项目已经完成交付验收，公司委派大数据运维工程师王工前往客户现场，为客户演示HDFS文件系统的查看与使用。

本项目以"指导客户使用大数据产品"为典型工作场景，针对"使用HDFS"的典型工作任务，完成两个子任务：使用HDFS的Web界面、使用Shell管理HDFS文件与目录。

在开展任务前，需要掌握必要的理论知识：什么是HDFS？其内部组件有哪些？各组件负责哪些职能？如何使用Shell实现HDFS文件系统操作？

完成任务后，进一步学习HDFS体系结构及其数据读写过程，加深对HDFS的认识，有助于实现对HDFS的排错、优化。

学习目标

1. 了解HDFS简介与设计目标。
2. 了解HDFS基本概念。
3. 熟练掌握HDFS文件系统的命令操作。
4. 了解HDFS的Web界面。
5. 熟悉HDFS体系结构。
6. 熟悉HDFS文件系统读写流程。

思维导图

项目思维导图如图4-1所示。

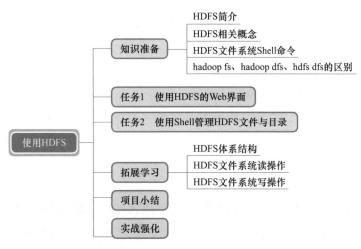

图4-1 项目思维导图

知识准备

1. HDFS简介

Hadoop分布式文件系统（Hadoop Distributed File System，HDFS）与现有的分布式文件系统类似，具有高度容错性、高吞吐量访问，适用于拥有大量数据集的应用程序。最初是作为Apache Nutch网络搜索引擎项目的基础设施而构建的，是Apache Hadoop Core项目的一部分。

HDFS的设计目标是希望HDFS运行在以下场景，见表4-1。

表4-1 HFDS设计目标

设 计 目 标	说　　明
硬件故障	HDFS实例运行在成百上千台廉价服务器集群上，每台服务器都存储部分文件系统数据，检测故障并从中快速自动恢复是HDFS的核心架构目标
流数据访问	HDFS更多适用于批处理场景，要求实现数据的高吞吐量
大型数据集	HDFS支持大文件，其典型文件大小为GB到TB，提供高数据带宽，单个实例支持数千万个文件
简单文件模型	HDFS的文件模型为"一次写入，多次读取"。一个文件一旦创建、写入和关闭就不再更改，仅支持将内容附加到文件末尾，但不能在任意点更新。这种文件模型解决了数据一致性问题并实现了高吞吐量数据访问
"移动计算"而非"移动数据"	将计算迁移到更靠近数据所在的位置通常比将数据移动到应用程序运行的位置更好。最大限度地减少了网络拥塞并增加了系统的整体吞吐量
可移植性	HDFS可以轻松地从一个平台移植到另一个平台，使HDFS可以应用于大量应用程序

2. HDFS相关概念

HDFS集群启动后，系统会有三种名称的进程，分别是NameNode、Secondary NameNode、DataNode，对应HDFS体系结构中的三个重要角色。

（1）名称节点（NameNode）

名称节点负责管理分布式文件系统的命名空间、文件与文件夹的元数据、数据块所在的数据节点位置信息。

针对文件系统的命名空间，名称节点保存了两个核心数据结构，即FsImage和EditLog。FsImage用来维护文件系统树以及文件树中所有的文件和文件夹的元数据。EditLog中记录了所有针对文件的创建、删除、重命名等操作。

名称节点在系统启动时，扫描所有数据节点，获取数据块所在的数据节点位置信息，并存在内存中。

另外，名称节点还控制客户端对文件的访问。

（2）第二名称节点（Secondary NameNode）

第二名称节点具有两大功能：首先，负责定时对名称节点中的FsImage和EditLog文件进行合并操作，减少EditLog的文件大小。其次，作为名称节点的"检查点"，保存名称节点中的系统命名空间信息，为集群提供"冷备份"功能。

（3）数据节点（DataNode）

数据节点负责数据的存储和读取，根据客户端或名称节点的调度来进行数据的存储和检索，定时向名称节点发送自己所存储的块的列表信息。数据节点需要将HDFS中的数据块存放于本地Linux文件系统中。

3. HDFS文件系统Shell命令

HDFS是以文件和目录的形式组织用户数据。它提供了一个名为FS Shell的命令行界面，允许用户与HDFS中的数据进行交互，实现文档的上传、下载、复制、查看文件信息、格式化名称节点等操作。该命令集的语法类似于用户已经熟悉的其他Shell（例如bash、csh）。常用操作见表4-2。

扫码看视频　　扫码看视频

表4-2　HDFS文件系统Shell命令

命令	使用方法	功　能
ls	hdfs dfs -ls 路径名称	列出HDFS文件系统指定目录下的目录和文件
cat	hdfs dfs -cat 文件名	将路径指定文件的内容输出到stdout
mkdir	hdfs dfs -mkdir 路径名称	接受路径指定的URI作为参数，创建这些目录
touchz	hdfs dfs -touchz 文件名	创建一个0字节的空文件
cp	hdfs dfs -cp 文件名　文件名	复制文件
mv	hdfs dfs -mv 文件或路径　文件或路径	移动文件
rm	hdfs dfs -rm 文件或路径	删除指定的文件
chgrp	hdfs dfs -chgrp [-R] 组名　文件或路径	改变文件所属的组。使用-R　将使改变在目录结构下递归进行
chmod	hdfs dfs -chmod [-R] 文件或路径	改变文件的权限。使用-R　将使改变在目录结构下递归进行
chown	hdfs dfs -chown [-R] [用户名]　文件或路径	改变文件的拥有者。使用-R　将使改变在目录结构下递归进行
get	hdfs dfs -get HDFS 文件名　本地文件名或路径	复制HDFS文件至本地文件系统
put	hdfs dfs -put 本地文件名　HDFS文件名或路径	复制本地文件至HDFS文件系统

4. hadoop fs、hadoop dfs、hdfs dfs的区别

Hadoop支持很多Shell命令，比如hadoop fs、hadoop dfs和hdfs dfs都是HDFS最常用的Shell命令，用来查看HDFS文件系统的目录结构、上传和下载数据、创建文件等。其适用场景及区别见表4-3。

表4-3　hadoop fs、hadoop dfs、hdfs dfs的适用场景及区别

HDFS命令	适用于HDFS	适用于本地文件系统
hadoop fs	√	√
hadoop dfs	√	×
hdfs dfs	√	×

任务1　使用HDFS的Web界面

扫码看视频

任务描述

客户想通过可视化的界面查看分布式文件系统HDFS的各项信息以及文件夹目录结构，由王工为其演示。

任务分析

1. 任务目标

1）能正确打开HDFS的Web界面。

2）了解HDFS的Web界面的常用功能界面。

3）熟练使用HDFS的Web界面查看文件系统信息。

2. 任务环境

操作系统：CentOS Stream 9（桌面环境）

预装软件版本：Java 1.8.0、Hadoop 3.3.4（伪分布模式/全分布模式）

3. 任务导图

任务导图如图4-2所示。

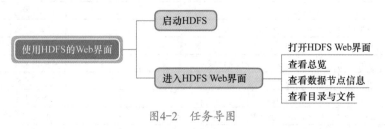

图4-2 任务导图

任务实施

1. 启动HDFS

使用hadoop用户启动HDFS。

```
$ start-dfs.sh
```

运行jps检查HDFS是否正常启动，如图4-3所示。

```
$ jps
```

```
[hadoop@localhost hadoop]$ start-dfs.sh
Starting namenodes on [localhost]
Starting datanodes
Starting secondary namenodes [localhost.localdomain]
[hadoop@localhost hadoop]$ jps
5749 SecondaryNameNode
5322 NameNode
5482 DataNode
5900 Jps
```

图4-3 启动HDFS

当看到NameNode、DataNode、SecondaryNameNode，说明HDFS已正常启动。

如果发现缺少某一个进程，怎么办？

1）尝试重启HDFS。

2）在HDFS上没有重要文件的前提下，可以尝试重新格式化HDFS：

① 停止HDFS；

② 删除/usr/local/hadoop/tmp文件；

③ 格式化HDFS；

④ 启动HDFS。

⚠ **注意**

启动HDFS后，本任务环境将存在两个文件系统：本地文件系统、HDFS分布式文件系统。读者要明确自己想要操作的是哪个文件系统，以使用相应指令。

⚠ **注意**

HDFS刚启动时，会有一段时间处于安全模式，并提示：NameNode is in safe mode。该模式下，不允许用户进行任何修改文件的操作，包括上传、删除、重命名、创建等。正常情况下，只需等待一段时间，HDFS会自动结束安全模式，用户便可正常读写了。

2. 进入HDFS Web界面

（1）打开HDFS Web界面

打开浏览器，如图4-4所示。

图4-4 打开浏览器

在地址栏输入HDFS Web地址，就可以看到HDFS的Web管理界面，如图4-5所示。

图4-5 访问HDFS的Web界面

（2）查看总览

HDFS的Web界面中的总览内容如图4-6所示。

> ⚠ 注意
>
> 本任务需要在桌面环境中进行。

> 📝 笔记
>
> 如果没有桌面环境，可以root用户使用指令安装：
> # yum groupinstall -y "GNOME Desktop"
> 安装完成后，使用指令打开桌面环境。
> # init 5

> 📝 笔记
>
> 如果操作系统还没有安装浏览器，可登录root用户，使用指令安装firefox浏览器。
> $ su - root
> # yum install -y firefox

> 📝 笔记
>
> 在Hadoop 2.x版本中，NameNode的默认访问端口为50070。从Hadoop 3.x版本起，端口改为9870。

> 📝 说明
>
> 用同样的方式，可以访问伪分布模式Hadoop的HDFS Web界面。

图4-6　HDFS的Web界面——总览

（3）查看数据节点信息

单击Datanodes，查看数据节点信息，如图4-7所示。

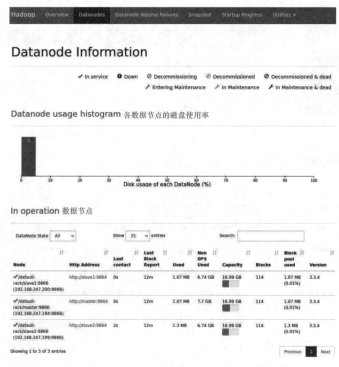

图4-7　HDFS的Web界面——数据节点信息

（4）查看目录与文件

单击utilities→Browse the file system，查看目录与文件，如图4-8所示。

如果在刚完成初始化后进入该页面，将没有任何文件。可以在后续任务过程中再进来查看。

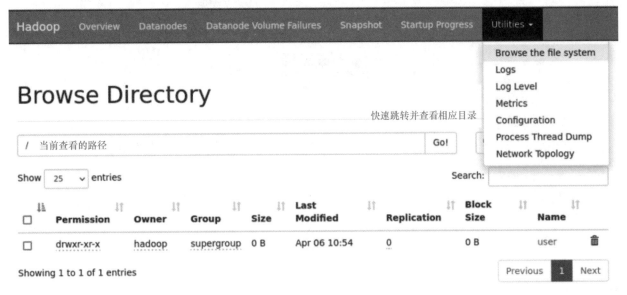

图4-8　HDFS的Web界面——目录与文件

任务2　使用Shell管理HDFS文件与目录

扫码看视频

任务描述

客户公司有一位略懂技术的工作人员，希望日常能简单地自行维护分布式文件系统HDFS上的文件，希望王工对其做常用指令的指导。王工依次向其演示查看目录、创建目录、删除目录、本地文件上传至HDFS、HDFS文件下载至本地、移动与复制HDFS文件。

任务分析

1. 任务目标

1）能理解创建用户目录的意义。

2）熟练使用Shell命令在HDFS文件系统中实现目录基本操作：创建、查看、删除。

3）熟练使用Shell命令实现本地文件与HDFS文件的交互操作：上传文件至HDFS、HDFS文件下载至本地。

4）熟练使用Shell命令实现HDFS文件操作，例如文件移动与复制。

2. 任务环境

操作系统：CentOS Stream 9

预装软件版本：Java 1.8.0、Hadoop 3.3.4（伪分布模式/全分布模式）

3. 任务导图

任务导图如图4-9所示。

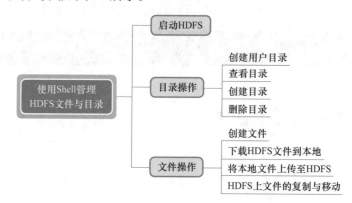

图4-9　任务导图

任务实施

1. 启动HDFS

使用hadoop用户启动HDFS。

```
$ start-dfs.sh
```

运行jps检查HDFS是否正常启动，如图4-10所示。

```
$ jps
```

```
[hadoop@localhost hadoop]$ start-dfs.sh
Starting namenodes on [localhost]
Starting datanodes
Starting secondary namenodes [localhost.localdomain]
[hadoop@localhost hadoop]$ jps
5749 SecondaryNameNode
5322 NameNode
5482 DataNode
5900 Jps
```

图4-10　启动HDFS

当看到NameNode、DataNode、SecondaryNameNode，说明HDFS已正常启动。

如果发现缺少某一个节点类型，按照任务1中的方式，重启HDFS或重新格式化HDFS。

2. 目录操作

（1）创建用户目录

本任务采用hadoop用户实现对HDFS文件系统的操作，因此需要在HDFS中为hadoop用户创建一个用户目录，如图4-11所示。

```
$ hdfs dfs -mkdir -p /user/hadoop
```

⚠ **注意**

启动HDFS后，本任务环境将存在两个文件系统：本地文件系统、HDFS分布式文件系统。读者要明确自己想要操作的是哪个文件系统，以使用相应指令。

⚠ **注意**

HDFS刚启动时，会有一段时间处于安全模式，并提示：NameNode is in safe mode。该模式下，不允许用户进行任何修改文件的操作，包括上传、删除、重命名、创建等。正常情况下，只需等待一段时间，HDFS会自动结束安全模式，用户便可正常读写了。

📝 **笔记**

hdfs dfs 命令的统一格式是类似"hdfs dfs -ls"这种形式，即在"-"后面跟上具体的操作。

项目4 使用HDFS

```
[hadoop@localhost hadoop]$ hdfs dfs -mkdir -p /user/hadoop
[hadoop@localhost hadoop]$ hdfs dfs -ls /
Found 1 items
drwxr-xr-x   - hadoop supergroup          0 2022-10-04 07:30 /user
[hadoop@localhost hadoop]$ hdfs dfs -ls /user
Found 1 items
drwxr-xr-x   - hadoop supergroup          0 2022-10-04 07:30 /user/hadoop
```

图4-11 创建用户目录

> **笔记**
> Hadoop 系统安装好以后，第一次使用 HDFS 时，需要首先在 HDFS 中创建用户目录，例如 hadoop 用户的用户目录为 /user/hadoop。

（2）查看目录

使用相对路径查看用户目录下的内容，如图4-12所示。

```
$ hdfs dfs -ls .
```

```
[hadoop@localhost hadoop]$ hdfs dfs -ls .
[hadoop@localhost hadoop]$
```

图4-12 查看用户目录

> **笔记**
> 在 HDFS 上进行文件操作时，默认当前路径"."为用户目录。

使用绝对路径查看HDFS的所有目录，如图4-13所示。

```
$ hdfs dfs -ls /
```

```
[hadoop@localhost hadoop]$ hdfs dfs -ls /
Found 1 items
drwxr-xr-x   - hadoop supergroup          0 2022-10-04 07:34 /user
```

图4-13 查看HDFS根目录

> **注意**
> 如果系统中不存在用户目录，则会出现错误提示，如图 4-14 所示。
>
> ```
> [hadoop@localhost hadoop]$ hdfs dfs -ls .
> ls: `.': No such file or directory
> ```
>
> 图4-14 无用户目录的错误提示

（3）创建目录

在用户目录下创建一个input文件夹并查看，如图4-15所示。

```
$ hdfs dfs -mkdir input
$ hdfs dfs -ls .
```

```
[hadoop@localhost hadoop]$ hdfs dfs -mkdir input
[hadoop@localhost hadoop]$ hdfs dfs -ls .
Found 1 items
drwxr-xr-x   - hadoop supergroup          0 2022-10-04 07:37 input
```

图4-15 在用户目录创建input文件夹

> **注意**
> 与 Linux 文件系统相似，在 HDFS 文件系统中，注意区分相对路径与绝对路径。在 HDFS 中，所有相对路径表示形式都是相对用户目录（如 /user/hadoop）路径而言的。

在HDFS根目录下创建一个input文件夹并查看，如图4-16所示。

```
$ hdfs dfs -mkdir /input
$ hdfs dfs -ls /
```

```
[hadoop@localhost hadoop]$ hdfs dfs -mkdir /input
[hadoop@localhost hadoop]$ hdfs dfs -ls /
Found 2 items
drwxr-xr-x   - hadoop supergroup          0 2022-10-04 07:38 /input
drwxr-xr-x   - hadoop supergroup          0 2022-10-04 07:34 /user
```

图4-16 在HDFS根目录创建input文件夹

（4）删除目录

删除用户目录下的input文件夹，如图4-17所示。

```
$ hdfs dfs -rm -r input
```

```
[hadoop@localhost hadoop]$ hdfs dfs -rm -r input
Deleted input
[hadoop@localhost hadoop]$ hdfs dfs -ls .
```

图4-17 删除用户目录下的input文件夹

> **笔记**
> 在本地文件系统中，如果要删除目录，指令是：rm -r 目录名。"-r"表示删除目录及其子目录下的所有内容。相应地，在 HDFS 文件系统操作时，只需在本地指令前增加 hdfs dfs -，即 hdfs dfs -rm -r 目录名。

3. 文件操作

（1）创建文件

在HDFS上创建文件，如图4-18所示。

```
$ hdfs dfs -touchz MyHDFSFile.txt
```

```
[hadoop@localhost hadoop]$ hdfs dfs -ls .
[hadoop@localhost hadoop]$ hdfs dfs -touchz MyHDFSFile.txt
[hadoop@localhost hadoop]$ hdfs dfs -ls .
Found 1 items
-rw-r--r--   1 hadoop supergroup          0 2022-10-04 07:42 MyHDFSFile.t
xt
```

图4-18　在HDFS上创建文件

（2）下载HDFS文件到本地

将HDFS文件系统中用户目录下的MyHDFSFile.txt文件下载到本地文件系统中，下载得到的文件放在本地文件系统的~/Downloads路径下，如图4-19所示。

```
$ mkdir ~/Downloads
$ hdfs dfs -get MyHDFSFile.txt ~/Downloads
```

查看从HDFS上下载得到的文件内容：

```
$ cat ~/Downloads/MyHDFSFile.txt
```

```
[hadoop@localhost ~]$ mkdir ~/Downloads
[hadoop@localhost ~]$ hdfs dfs -get MyHDFSFile.txt ~/Downloads/
[hadoop@localhost ~]$ cat ~/Downloads/MyHDFSFile.txt
[hadoop@localhost ~]$ ll ~/Downloads/MyHDFSFile.txt
-rw-r--r--. 1 hadoop hadoop 0 10月  4 08:03 /home/hadoop/Downloads/MyHDFSFile.txt
```

图4-19　下载HDFS文件到本地

（3）将本地文件上传至HDFS

编辑本地MyHDFSFile.txt文件。

```
$ vim ~/Downloads/MyHDFSFile.txt
```

编辑文件内容，例如：

```
This is my HDFS file.
I edit this file in the local file system.
```

保存并退出，如图4-20所示。

```
[hadoop@localhost ~]$ vim ~/Downloads/MyHDFSFile.txt
[hadoop@localhost ~]$ cat ~/Downloads/MyHDFSFile.txt
This is my HDFS file.
I edit this file in the local file system.
```

图4-20　在本地创建文件

将MyHDFSFile.txt文件上传到HDFS文件系统中的用户目录下，并覆盖原同名文件，如图4-21所示，上传后，文件大小及日期信息发生改变。

说明

在使用HDFS文件时，如果有文本编辑需求，一般不会直接在HDFS上编辑，而是先将文件下载到本地文件系统，并在本地完成文本编辑后重新上传HDFS。

笔记

复制HDFS文件至本地文件系统：

hdfs dfs -get HDFS文件名 本地文件名或路径

```
$ hdfs dfs -put -f ~/Downloads/MyHDFSFile.txt .
$ hdfs dfs -ls .
$ hdfs dfs -cat MyHDFSFile.txt
```

```
[hadoop@localhost ~]$ hdfs dfs -ls .
Found 1 items
-rw-r--r--   1 hadoop supergroup          0 2022-10-04 07:42 MyHDFSFile.txt
[hadoop@localhost ~]$ hdfs dfs -put -f ~/Downloads/MyHDFSFile.txt .
[hadoop@localhost ~]$ hdfs dfs -ls .
Found 1 items
-rw-r--r--   1 hadoop supergroup         65 2022-10-04 08:17 MyHDFSFile.txt
[hadoop@localhost ~]$ hdfs dfs -cat MyHDFSFile.txt
This is my HDFS file.
I edit this file in the local file system.
```

图4-21　将本地文件上传到HDFS文件系统中

笔记

复制本地文件至HDFS文件系统：
hdfs dfs -put 本地文件名 HDFS 文件名或路径

-f表示如遇到名称冲突则覆盖原文件。

（4）HDFS上文件的复制与移动

复制MyHDFSFile.txt文件为MyHDFSFile.txt.bak，对文件进行备份，如图4-22所示。

```
$ hdfs dfs -cp MyHDFSFile.txt MyHDFSFile.txt.bak
$ hdfs dfs -ls .
```

```
[hadoop@localhost ~]$ hdfs dfs -ls .
Found 1 items
-rw-r--r--   1 hadoop supergroup         65 2022-10-04 08:17 MyHDFSFile.txt
[hadoop@localhost ~]$ hdfs dfs -cp MyHDFSFile.txt MyHDFSFile.txt.bak
[hadoop@localhost ~]$ hdfs dfs -ls .
Found 2 items
-rw-r--r--   1 hadoop supergroup         65 2022-10-04 08:17 MyHDFSFile.txt
-rw-r--r--   1 hadoop supergroup         65 2022-10-04 08:21 MyHDFSFile.txt.bak
```

图4-22　HDFS上复制文件

笔记

复制文件：
hdfs dfs -cp 文件名 文件名

在用户目录下创建名为bak的备份文件夹，用来存储备份文件MyHDFSFile.txt.bak。创建文件夹./bak，并移动MyHDFSFile.txt.bak到./bak中，如图4-23所示。

```
$ hdfs dfs -mkdir ./bak
$ hdfs dfs -mv MyHDFSFile.txt.bak ./bak
```

```
[hadoop@localhost ~]$ hdfs dfs -ls .
Found 2 items
-rw-r--r--   1 hadoop supergroup         65 2022-10-04 08:17 MyHDFSFile.txt
-rw-r--r--   1 hadoop supergroup         65 2022-10-04 08:21 MyHDFSFile.txt.bak
[hadoop@localhost ~]$ hdfs dfs -mkdir ./bak
[hadoop@localhost ~]$ hdfs dfs -ls .
Found 3 items
-rw-r--r--   1 hadoop supergroup         65 2022-10-04 08:17 MyHDFSFile.txt
-rw-r--r--   1 hadoop supergroup         65 2022-10-04 08:21 MyHDFSFile.txt.bak
drwxr-xr-x   - hadoop supergroup          0 2022-10-04 08:24 bak
[hadoop@localhost ~]$ hdfs dfs -mv MyHDFSFile.txt.bak ./bak
[hadoop@localhost ~]$ hdfs dfs -ls .
Found 2 items
-rw-r--r--   1 hadoop supergroup         65 2022-10-04 08:17 MyHDFSFile.txt
drwxr-xr-x   - hadoop supergroup          0 2022-10-04 08:25 bak
[hadoop@localhost ~]$ hdfs dfs -ls ./bak
Found 1 items
-rw-r--r--   1 hadoop supergroup         65 2022-10-04 08:21 bak/MyHDFSFile.txt.bak
```

图4-23　HDFS上移动文件

笔记

移动文件：
hdfs dfs -mv 文件或路径 文件或路径

拓展学习

1. HDFS体系结构

HDFS是主/从架构,由单个名称节点和N个数据节点组成,如图4-24所示。

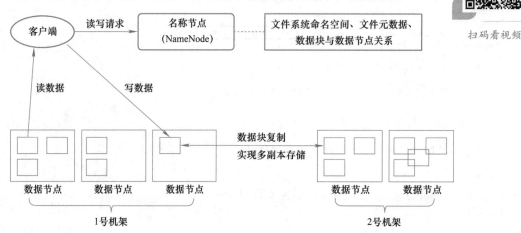

图4-24　HDFS体系结构图

HDFS文件系统有一个命名空间,允许将用户数据存储在文件中,每个文件被分成一个或多个块,这些块存储在一组数据节点中。

名称节点执行文件系统命名空间操作,例如,打开、关闭和重命名文件和目录,同时确定数据块到数据节点的映射。

数据节点负责处理来自文件系统客户端的读取和写入请求,执行块创建、删除。

集群中单个名称节点的设计方式极大地简化了系统的架构。名称节点是所有HDFS元数据的仲裁者和存储库。该系统的设计方式使用户数据永远不会流经名称节点,避免了让名称节点成为数据传输的瓶颈。

2. HDFS文件系统读操作

HDFS文件系统读操作如图4-25所示。

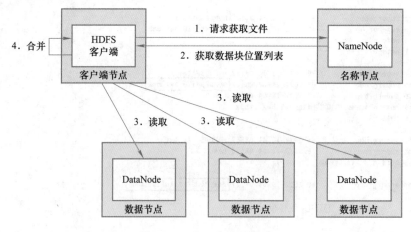

图4-25　HDFS文件读操作

1）客户端向名称节点发起请求，通过RPC来调用名称节点，请求获取文件的数据块所在位置信息。

2）名称节点返回数据块的数据节点地址。这些数据节点根据它们与客户端的距离自动排序。

3）客户端按照数据块的数据节点列表，按序访问数据节点，并对每一个获取得到的数据块进行完整性校验，若数据块不完整，则向列表中的下一个数据节点获取该数据块。

4）客户端将所有数据块合并成一个完整的最终文件。

3. HDFS文件系统写操作

HDFS文件系统写操作如图4-26所示。

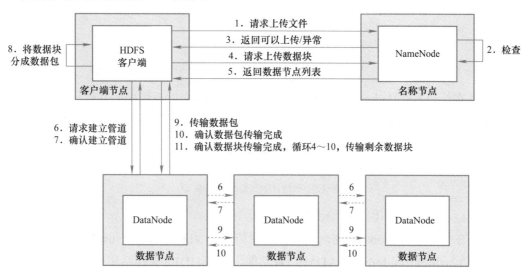

图4-26　HDFS文件写操作

1）客户端使用RPC调用名称节点，请求上传文件。

2）名称节点检查系统目录树，确保这个文件不存在且客户端具有在HDFS上创建该文件的权限。

3）如果检测通过，名称节点就会生成一个新的文件记录，并告知客户端可以上传文件；否则，文件创建失败，并向客户端抛出一个异常。

4）客户端请求上传第一个数据块及数据块数量。数据块数量可以自定义，也可以按照集群规划的数量（默认为3个副本）。

5）名称节点返回可以用于存储该数据块的数据节点的IP列表。

6）客户端接收数据节点的IP列表后，访问其中一个数据节点，请求建立管道，第一个数据节点收到请求后，向第二个数据节点发送请求，第二个数据节点接收到请求后向第三个数据节点发送请求。

7）由第三个数据节点开始，依次返回确认信息，完成管道建立。

8）客户端将数据块分成一个个的数据包（默认64KB），写入内部的队列，成为数据队列。

9）客户端将数据包分流给管道中的第一个数据节点，这个节点会存储数据包并且发送给管道中的第二个数据节点。同样地，第二个数据节点存储数据包并且传给管道中的第三个数据节点。

10）第三个数据节点收到数据包后，反向向第二个数据节点发送ACK应答，确认收到数据包。ACK应答随后发送至第一个数据节点，最后发送给客户端。

11）第一个数据块中的所有数据包发送完成后，客户端发起第二个数据块发送请求，并循环4）～10）流程，直到文件的所有数据块发送完成。

项目小结

本项目以"指导客户使用大数据产品"为典型工作场景，针对使用HDFS的典型工作任务，开展围绕HDFS的理论与任务学习。在了解了HDFS相关概念、Shell指令的基础上，进一步学习使用HDFS，介绍了两种常用的HDFS使用方法：Web界面以及基于Shell的目录文件操作。完成任务后，继续学习HDFS体系结构，了解HDFS分布式文件系统的内部运行机制，这些有利于对HDFS开展排错及性能优化工作。

实战强化

1. 运维人员在伪分布式Hadoop部署环境下，想删除HDFS文件系统中的文件MyHDFSFile.txt，使用命令hdfs dfs -rm MyHDFSFile.txt，得到的结果如图4-27所示。

```
[hadoop@localhost ~]$ hdfs dfs -rm MyHDFSFile.txt
rm: Cannot delete /user/hadoop/MyHDFSFile.txt. Name node is in safe mode.
```

图4-27 删除HDFS文件错误提示

试分析该运维人员出错的原因，并给出解决方案。

2. 运维人员在伪分布式Hadoop部署环境下，使用传输命令hdfs dfs -put /usr/local/hadoop/README.txt，实现将本地/usr/local/hadoop/README.txt文件上传到HDFS文件系统中，系统响应如图4-28所示。

```
[hadoop@localhost ~]$ hdfs dfs -put /usr/local/hadoop/README.txt .
put: `.': No such file_or directory: `hdfs://localhost:9000/user/hadoop'
```

图4-28 传输文件到HDFS错误提示

请问在该命令中，运维人员会将文件上传到HDFS的什么位置，试分析其出错的原因，并给出解决方案。

项目 5　MapReduce 编程

● 项目概述

A公司根据客户的大数据业务场景需求，开发完成了一套基于Hadoop的大数据平台项目。项目进入交付验收阶段，公司的大数据运维工程师王工已经完成了基于Hadoop的集群部署。为了验证集群有效性，同时学习MapReduce的编程思想，王工决定编写MapReduce典型实例——词频统计，并在集群中进行功能验证。

本项目针对"MapReduce编程"的典型工作任务，编写程序实现词频统计功能。

在开展任务前，需要掌握必要的理论知识：什么是MapReduce？在Hadoop的版本变迁中，MapReduce与YARN的关系是怎样的？MapReduce的核心思想、设计理念、体系结构、编程思想分别是怎样的？再学习词频统计的实例，便于开展MapReduce编程设计。

完成任务后，进一步了解与MapReduce相关的知识，加深对MapReduce的理解。

● 学习目标

1. 了解MapReduce与YARN的关系。
2. 理解MapReduce核心思想与设计理念。
3. 了解MapReduce体系结构。
4. 理解MapReduce编程思想。
5. 会编写、执行MapReduce版本WordCount。
6. 了解MapReduce工作流程。
7. 能简述WordCount的具体执行过程。

● 思维导图

项目思维导图如图5-1所示。

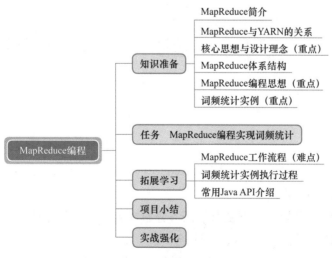

图5-1　项目思维导图

知识准备

1. MapReduce简介

MapReduce是一种基于Java语言开发的分布式计算框架，支持多种编程语言。核心功能是将用户编写的业务逻辑代码和自带默认组件整合成一个完整的分布式运算程序，并发布运行在一个Hadoop集群上。MapReduce计算框架适用于批处理任务，即在可接受的时间内对整个数据集计算某个特定的查询结果，该计算模型不适合需要实时反映数据变化状态的计算环境。

使用计算框架的意义则是使并行计算开发经验较少的人员也能轻松地实现并行计算应用程序。

2. MapReduce与YARN的关系

Hadoop 1.0到Hadoop 2.0，其核心组件与组件功能如图5-2所示。

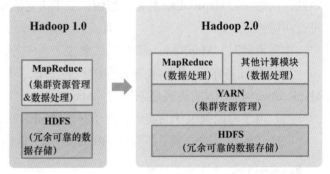

图5-2 Hadoop 1.0到Hadoop 2.0的核心组件与组件功能

在Hadoop 1.0版本中，核心角色仅有HDFS与MapReduce，HDFS负责数据的存储，MapReduce需要同时负责数据计算及计算资源的管理调度。因此，在Hadoop 1.0中存在资源管理效率低的缺点，基于这样的缺点，在Hadoop 2.0中，增加了新的资源管理框架YARN。从Hadoop 2.0起，MapReduce仅负责并行计算，是一个纯粹的计算框架，而资源管理相关任务全部交由YARN，YARN是一个纯粹的资源管理调度框架。

在Hadoop中，YARN作为一个资源管理调度框架，是Hadoop下MapReduce程序运行的生存环境。但同时，MapRuduce除了可以运行YARN框架下，也可以运行在诸如Mesos、Corona之类的调度框架上。使用不同的调度框架，需要针对Hadoop做不同的适配。

3. 核心思想与设计理念

（1）核心思想——分而治之

MapReduce核心思想为"分而治之"，把一个大规模的数据集切分成很多小的单独的数据集，然后放在多个机器上并行处理，如图5-3所示。

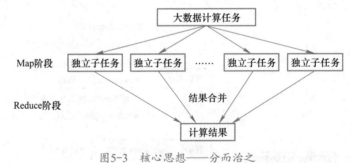

图5-3 核心思想——分而治之

使用MapReduce操作海量数据时，每个MapReduce程序被初始化为一个大数据计算任务，每个任务可以分为Map阶段和Reduce阶段。

Map阶段：负责将任务分解，即把复杂的任务分解成若干个"简单的独立子任务"来并行处理，这些任务没有必然的依赖关系，可以单独执行任务。

Reduce阶段：负责将结果合并，即把Map阶段的结果进行全局汇总。

这样做的好处是可以在任务被分解后，将数据交给不同的机器去处理，实现并行计算，减少整个操作的时间。因此，"MapReduce"即"分解+合并"。

在"分而治之"的过程中，不同的Map任务之间不会进行通信，不同的Reduce任务之间也不会发生任何信息交换，用户不能显式地从一台机器向另一台机器发送消息，所有的数据交换都是通过MapReduce框架自身去实现的。

（2）设计理念——计算向数据靠拢

由于在集群中，移动数据需要大量的网络传输开销，因此，MapReduce的设计理念是"计算向数据靠拢"，而不是"数据向计算靠拢"。图5-4分别展示了"数据向计算靠拢"与"计算向数据靠拢"情况下的集群。通过分发应用程序，避免大量数据网络传输带来的延迟。

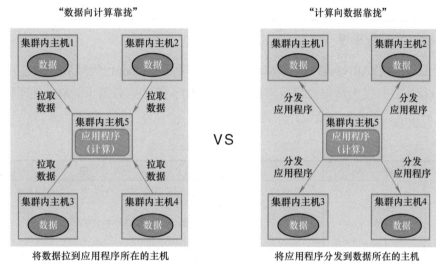

图5-4　数据向计算靠拢vs计算向数据靠拢

因此，在Map阶段，应用程序会被分发到数据所在的节点上，并执行map任务。

4. MapReduce体系结构

MapReduce框架采用了Master/Slave架构，包括一个Master和若干个Slave。在Master上运行JobTracker，在Slave上运行TaskTracker。体系结构如图5-5所示。

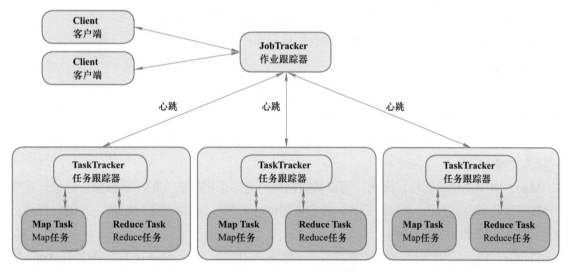

图5-5　MapReduce体系结构

体系结构图中的各部分功能介绍见表5-1。

表5-1　MapReduce各部分功能介绍

名　称	功　能
客户端 （Client）	1）编写、提交MapReduce程序； 2）通过接口查询当前应用程序（作业）的运行状态
作业跟踪器 （JobTracker）	1）监控资源使用量； 2）初始化作业、分配作业； 3）监控各任务跟踪器（TaskTracker）的健康状况； 4）监控当前应用程序（作业）的运行状态； 5）一旦检测到失败，则把任务转移到其他节点继续执行
任务跟踪器 （TaskTracker）	1）接收作业跟踪器（JobTracker）发送的命令，执行具体的任务； 2）将自身的资源使用情况、任务执行进度，通过心跳的方式，发送给作业跟踪器（JobTracker）

5. MapReduce编程思想

在Hadoop中编写依赖YARN框架执行的MapReduce程序时，Hadoop已经默认提供通用的YARNRunner和MRAppMaster程序，开发人员仅需编写满足接口要求的、符合业务需求的Map处理和Reduce处理程序即可。其中，Map函数和Reduce函数介绍见表5-2。

表5-2　Map函数与Reduce函数

函　数	输　入	输　出	说　明
Map	<k1,v1> 如：<行号,"a b c">	List(<k2,v2>) 如："a",1 <"b",1> <"c",1>	1）将小数据集进一步解析成一批<key,value>对，输入Map函数中进行处理 2）每一个输入的<k1,v1>会输出一批<k2,v2>。<k2,v2>是计算的中间结果
Reduce	<k2,List(v2)> 如：<"a",<1,1,1>>	<k3,v3> 如：<"a",3>	输入的中间结果<k2,List(v2)>中的List(v2)表示是一批属于同一个k2的value

编写一个MapReduce程序并不复杂，关键点在于掌握分布式的编程思想和方法，主要将计算过程分为以下五个步骤：

1）输入。遍历输入数据，并将之解析成键值对<k1,v1>。

2）Map。将输入<k1,v1>映射（map）成另外一些键值对，即List(<k2,v2>)。

3）Shuffle。依据k2值，对中间数据进行分组（group），生成<k2,List(v2)>。

4）Reduce。以k2为单位对数据进行归约（reduce），产生新键值对<k3,v3>。

5）输出。将最终产生的键值对<k3,v3>保存到输出文件中。

6. 词频统计实例

词频统计是指对输入的文本文件进行分析，输出文件中每个单词及其出现的次数（频数）。图5-6展示的是对两个文本文件进行词频统计的过程。

1）输入：通过默认组件TextInputFormat将内容转换为<key,value>键值对，其中，key为行号，value为相应行的文本内容。

2）Map：图中文档共四行，因此，可以将数据集分为四个相互独立的子数据集，分别开展独立子任务。Map阶段，每行分配一个Map任务，分别调用map()方法，将单词进行切割，并进行计数，输出键值对列表List(<k2,v2>)。

3）Shuffle：将具有相同key的键值对归并成一个键值对，比如，两个<Hadoop,1>归并成<Hadoop,<1,1>>。在进行归并的同时，还会对结果进行排序。

4) Reduce：上述归并的结果会作为Reduce的输入，调用Reduce()方法，计算每个单词总的出现次数。

5) 输出：通过TextOutputFormat组件将Reduce的结果值输出到结果文件中。

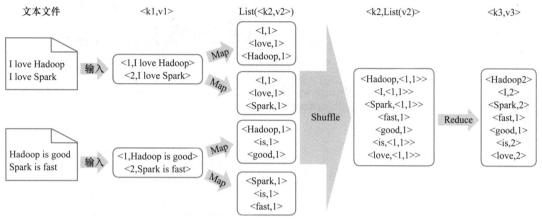

图5-6 WordCount.实例

任务 MapReduce编程实现词频统计

扫码看视频

任务描述

词频统计指的是对一系列文本文件中出现的所有单词进行分析、统计，输出各单词出现的次数。词频统计是MapReduce最典型的编程实例，就像"Hello World"在编程语言中的地位。通过基于MapReduce的编程实践，可以更好地理解MapReduce编程思想，有利于后续的MapReduce排错、调优工作。

任务分析

1. 任务目标

1) 理解WordCount案例的执行过程；
2) 基本理解WordCount案例的编程思路，包括map方法、reduce方法及主函数；
3) 了解常用的MapReduce Java API；
4) 熟练掌握jar包的打包、运行过程。

2. 任务环境

操作系统：CentOS Stream 9
软件版本：Java 1.8.0、Hadoop 3.3.4（伪分布模式/全分布模式）

3. 任务导图

任务导图如图5-7所示。

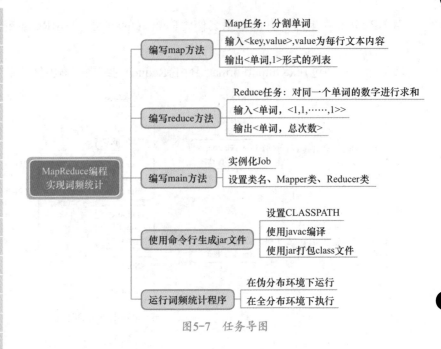

图5-7 任务导图

任务实施

1. 编写Map方法

本任务中，Map需要实现的逻辑是分割单词，输出内容为<单词，1>。

在Map阶段，文件中的文本数据被读入，以<key,value>的形式提交给Map函数进行处理，其中，key是当前读取到的行号，value是当前读取到的行的内容。<key,value>提交给Map函数以后，就运行自定义的Map处理逻辑，对value进行处理，然后以特定的键值对的形式进行输出，这个输出将作为中间结果，继续提供给Reduce阶段作为输入数据。

以下是Map处理逻辑的具体代码：

```
public static class TokenizerMapper extends Mapper<Object, Text, Text, IntWritable> {//继承自Hadoop的类Mapper。
    private static final IntWritable one = new IntWritable(1);// 初始化对象one，初始化为1，表示该单词出现过
    private Text word = new Text();
    public TokenizerMapper() {
    }
    public void map(Object key, Text value, Mapper<Object, Text, Text, IntWritable>.Context context) throws IOException, InterruptedException {
        StringTokenizer itr = new StringTokenizer(value.toString());
        while(itr.hasMoreTokens()) {
            this.word.set(itr.nextToken());
            context.write(this.word, one);
        }
    }
}
```

> **说明**
>
> 本任务的目的是对读取的文本文件进行词频统计。

> **说明**
>
> Map 函数的输入参数中 key、value 即为 Map 阶段的输入 <key,value>。value 即为每行文本内容。
>
> 实例化 Tokenizer 类，并利用 nextToken() 方法依次拆分单词，每拆分出一个单词，便生成一个对象 < 单词，1>。
>
> context 作为 map 函数的输出对象，存放了所有 < 单词，1> 组成的中间结果列表。

2. 编写reduce方法

本任务中，Reduce需要实现的逻辑是对输入结果中的数字序列进行求和。

经过Shuffle阶段，Map阶段得到的中间结果分发给对应的Reduce任务处理。对于Reduce阶段而言，输入是<key,value-list>形式，比如<'Hadoop',<1,1>>。Reduce函数就是对输入中的value-list进行求和，得到词频统计结果。

以下是Reduce处理逻辑的具体代码：

```java
    public static class IntSumReducer extends Reducer<Text, IntWritable, Text, IntWritable> {//继承自Hadoop的类Reducer
        private IntWritable result = new IntWritable(); //定义result变量记录每个单词出现的次数
        public IntSumReducer() {
        }
        public void reduce(Text key, Iterable<IntWritable> values, Reducer<Text, IntWritable, Text, IntWritable>.Context context) throws IOException, InterruptedException {
            int sum = 0;
            IntWritable val;
            for(Iterator i$ = values.iterator(); i$.hasNext(); sum += val.get()) {//遍历Iterable容器，对数字进行累加
                val = (IntWritable)i$.next();
            }
            this.result.set(sum);
            context.write(key, this.result); // result记录单词key的累加结果，写入context对象中
        }
    }
```

> **说明**
> 上一步 Map 的输出结果中，IntWritable 变量经过 Shuffle 阶段后，变成了 Iterable 容器。

3. 编写main方法

（1）编写main方法

为了让TokenizerMapper类和IntSumReducer类能够协同工作，完成最终的词频统计任务，需要在主函数中通过Job类设置Hadoop程序运行时的环境变量，具体代码如下：

```java
    public static void main(String[] args) throws Exception {
        Configuration conf = new Configuration();
        String[] otherArgs = (new GenericOptionsParser(conf, args)).getRemainingArgs();
        //通过类Configuration获得程序运行时的参数情况，并存储于String[] otherArgs
        if(otherArgs.length < 2) {//这里定义了调用该方法时的参数必须大于等于2，否则输出错误提示
            System.err.println("Usage: wordcount <in> [<in>...] <out>");
            System.exit(2);
        }
        Job job = Job.getInstance(conf, "word count");      //设置环境参数
        job.setJarByClass(WordCount.class);                 //设置整个程序的类名
        job.setMapperClass(WordCount.TokenizerMapper.class);//添加Mapper类
        job.setReducerClass(WordCount.IntSumReducer.class); //添加Reducer类
        job.setOutputKeyClass(Text.class);                  //设置输出类型
        job.setOutputValueClass(IntWritable.class);         //设置输出类型
        for(int i = 0; i < otherArgs.length - 1; ++i) {
```

```
            FileInputFormat.addInputPath(job, new Path(otherArgs[i]));  //设置
输入文件
        }
            FileOutputFormat.setOutputPath(job, new Path(otherArgs[otherArgs.
length - 1]));//设置输出文件
            System.exit(job.waitForCompletion(true)?0:1);
    }
```

（2）完整的词频统计程序

在编写词频统计Java程序时，需要新建一个名称为WordCount.java的文件，该文件包含了完整的词频统计程序代码，WordCount.java文件被放在了"/usr/local/hadoop"目录下。

```
$ cd /usr/local/hadoop
$ vim WordCount.java
```

具体代码如下：

```
import java.io.IOException;
import java.util.Iterator;
import java.util.StringTokenizer;
import org.apache.hadoop.conf.Configuration;
import org.apache.hadoop.fs.Path;
import org.apache.hadoop.io.IntWritable;
import org.apache.hadoop.io.Text;
import org.apache.hadoop.mapreduce.Job;
import org.apache.hadoop.mapreduce.Mapper;
import org.apache.hadoop.mapreduce.Reducer;
import org.apache.hadoop.mapreduce.lib.input.FileInputFormat;
import org.apache.hadoop.mapreduce.lib.output.FileOutputFormat;
import org.apache.hadoop.util.GenericOptionsParser;
public class WordCount {
    public WordCount() {
    }
    public static void main(String[] args) throws Exception {
        Configuration conf = new Configuration();
        String[] otherArgs = (new GenericOptionsParser(conf, args)).getRemainingArgs();
        if(otherArgs.length < 2) {
            System.err.println("Usage: wordcount <in> [<in>...] <out>");
            System.exit(2);
        }
        Job job = Job.getInstance(conf, "word count");
        job.setJarByClass(WordCount.class);
        job.setMapperClass(WordCount.TokenizerMapper.class);
        job.setCombinerClass(WordCount.IntSumReducer.class);
        job.setReducerClass(WordCount.IntSumReducer.class);
        job.setOutputKeyClass(Text.class);
        job.setOutputValueClass(IntWritable.class);
        for(int i = 0; i < otherArgs.length - 1; ++i) {
            FileInputFormat.addInputPath(job, new Path(otherArgs[i]));
        }
        FileOutputFormat.setOutputPath(job, new Path(otherArgs[otherArgs.length - 1]));
        System.exit(job.waitForCompletion(true)?0:1);
    }
    public static class TokenizerMapper extends Mapper<Object, Text, Text, IntWritable> {
```

```java
        private static final IntWritable one = new IntWritable(1);
        private Text word = new Text();
        public TokenizerMapper() {
        }
        public void map(Object key, Text value, Mapper<Object, Text, Text, IntWritable>.Context context) throws IOException, InterruptedException {
            StringTokenizer itr = new StringTokenizer(value.toString());
            while(itr.hasMoreTokens()) {
                this.word.set(itr.nextToken());
                context.write(this.word, one);
            }
        }
    }
    public static class IntSumReducer extends Reducer<Text, IntWritable, Text, IntWritable> {
        private IntWritable result = new IntWritable();
        public IntSumReducer() {
        }
        public void reduce(Text key, Iterable<IntWritable> values, Reducer<Text, IntWritable, Text, IntWritable>.Context context) throws IOException, InterruptedException {
            int sum = 0;
            IntWritable val;
            for(Iterator i$ = values.iterator(); i$.hasNext(); sum += val.get()) {
                val = (IntWritable)i$.next();
            }
            this.result.set(sum);
            context.write(key, this.result);
        }
    }
}
```

4. 使用命令行生成jar文件

把Hadoop安装目录设置为当前工作目录，命令如下：

```
$ cd /usr/local/hadoop
```

执行命令，设置CLASSPATH变量，该变量是为了让javac编译程序可以找到Hadoop相关的jar包，如图5-8所示。

```
$export CLASSPATH="/usr/local/hadoop/share/hadoop/common/hadoop-common-3.3.4.jar:/usr/local/hadoop/share/hadoop/mapreduce/hadoop-mapreduce-client-core-3.3.4.jar:/usr/local/hadoop/share/hadoop/common/lib/commons-cli-1.2.jar:$CLASSPATH"
```

```
[hadoop@localhost hadoop]$ cd /usr/local/hadoop
[hadoop@localhost hadoop]$ export CLASSPATH="/usr/local/hadoop/share/h
adoop/common/hadoop-common-3.3.4.jar:/usr/local/hadoop/share/hadoop/ma
preduce/hadoop-mapreduce-client-core-3.3.4.jar:/usr/local/hadoop/share
/hadoop/common/lib/commons-cli-1.2.jar:$CLASSPATH"
[hadoop@localhost hadoop]$ echo $CLASSPATH
/usr/local/hadoop/share/hadoop/common/hadoop-common-3.3.4.jar:/usr/loc
al/hadoop/share/hadoop/mapreduce/hadoop-mapreduce-client-core-3.3.4.ja
r:/usr/local/hadoop/share/hadoop/common/lib/commons-cli-1.2.jar:
```

图5-8 找到Hadoop相关的jar包

执行javac命令编译程序，如图5-9所示。

```
$ javac WordCount.java
```

> **说明**
>
> 需要用到 Java JDK 中的编译工具。

> **提示**
>
> 编译之后，在文件夹下可以发现有三个".class"文件，这是Java的可执行文件。

```
[hadoop@localhost hadoop]$ cd /usr/local/hadoop
[hadoop@localhost hadoop]$ ls WordCount.java
WordCount.java
[hadoop@localhost hadoop]$ javac WordCount.java
[hadoop@localhost hadoop]$ ls
bin              LICENSE.txt      sbin
etc              logs             share
include          myapp            tmp
lib              NOTICE-binary    'WordCount$IntSumReducer.class'
libexec          NOTICE.txt       'WordCount$TokenizerMapper.class'
LICENSE-binary   output           WordCount.class
licenses-binary  README.txt       WordCount.java
```

图5-9　编译程序生成class文件

将class文件打包并命名为WordCount.jar，如图5-10所示。

```
$ jar -cvf WordCount.jar *.class
```

```
[hadoop@localhost hadoop]$ jar -cvf WordCount.jar *.class
已添加清单
正在添加: WordCount$IntSumReducer.class(输入 = 1744) (输出 = 741)(压缩了 57%)
正在添加: WordCount$TokenizerMapper.class(输入 = 1740) (输出 = 756)(压缩了 56%)
正在添加: WordCount.class(输入 = 1911) (输出 = 1043)(压缩了 45%)
[hadoop@localhost hadoop]$ ls
bin              logs             tmp
etc              myapp            'WordCount$IntSumReducer.class'
include          NOTICE-binary    'WordCount$TokenizerMapper.class'
lib              NOTICE.txt       WordCount.class
libexec          output           WordCount.jar
LICENSE-binary   README.txt       WordCount.java
licenses-binary  sbin
LICENSE.txt      share
```

图5-10　打包生成jar文件

5. 运行词频统计程序

启动Hadoop。

```
$ start-dfs.sh
```

> **说明**
>
> 记得启动完后要用jps验证Hadoop各节点是否正常启动。

为了避免其他任务对本次任务的影响，先删除HDFS上/user/hadoop路径下的input文件夹和output文件夹。

```
$ hdfs dfs -rm -r input
$ hdfs dfs -rm -r output
```

删除后重新创建input文件夹，如图5-11所示。

```
$ hdfs dfs -mkdir input
```

```
[hadoop@localhost hadoop]$ hdfs dfs -ls .
[hadoop@localhost hadoop]$ hdfs dfs -mkdir input
[hadoop@localhost hadoop]$ hdfs dfs -ls .
Found 1 items
drwxr-xr-x   - hadoop supergroup          0 2022-08-09 02:47 input
```

图5-11　在HDFS用户目录中创建input文件夹

在本地文件系统中新建两个文件wordfile1.txt和wordfile2.txt。

```
$ cd /usr/local/hadoop
$ mkdir data
$ vim data/wordfile1.txt
$ vim data/wordfile2.txt
```

两个文本文件的内容如下：

wordfile1.txt：

I love Hadoop
I love Spark

wordfile2.txt：

Hadoop is good
Spark is fast

将两个文件上传到HDFS中的"/user/hadoop/input"目录下，如图5-12所示。

$ hdfs dfs -put /usr/local/hadoop/data/* input

```
[hadoop@localhost hadoop]$ cd /usr/local/hadoop
[hadoop@localhost hadoop]$ mkdir data
[hadoop@localhost hadoop]$ vim data/wordfile1.txt
[hadoop@localhost hadoop]$ vim data/wordfile2.txt
[hadoop@localhost hadoop]$ cat data/wordfile1.txt
I love Hadoop
I love Spark
[hadoop@localhost hadoop]$ cat data/wordfile2.txt
Hadoop is good
Spark is fast
[hadoop@localhost hadoop]$ hdfs dfs -put /usr/local/hadoop/data/* input
[hadoop@localhost hadoop]$ hdfs dfs -cat input/wordfile1.txt
I love Hadoop
I love Spark
[hadoop@localhost hadoop]$ hdfs dfs -cat input/wordfile2.txt
Hadoop is good
Spark is fast
```

图5-12　上传任务文件至HDFS

执行WordCount.jar，如图5-14所示。

$ cd /usr/local/hadoop
$ hadoop jar ./WordCount.jar WordCount input output

```
2022-10-05 22:23:49,072 INFO mapreduce.Job: Counters: 36
    File System Counters
        FILE: Number of bytes read=11613
        FILE: Number of bytes written=1927713
        FILE: Number of read operations=0
        FILE: Number of large read operations=0
        FILE: Number of write operations=0
        HDFS: Number of bytes read=141
        HDFS: Number of bytes written=47
        HDFS: Number of read operations=24
        HDFS: Number of large read operations=0
        HDFS: Number of write operations=5
        HDFS: Number of bytes read erasure-coded=0
    Map-Reduce Framework
        Map input records=4
        Map output records=12
        Map output bytes=104
        Map output materialized bytes=112
        Input split bytes=236
        Combine input records=12
        Combine output records=9
        Reduce input groups=7
        Reduce shuffle bytes=112
        Reduce input records=9
        Reduce output records=7
        Spilled Records=18
        Shuffled Maps =2
        Failed Shuffles=0
        Merged Map outputs=2
        GC time elapsed (ms)=15
        Total committed heap usage (bytes)=890765312
    Shuffle Errors
        BAD_ID=0
        CONNECTION=0
        IO_ERROR=0
        WRONG_LENGTH=0
        WRONG_MAP=0
        WRONG_REDUCE=0
    File Input Format Counters
        Bytes Read=56
    File Output Format Counters
        Bytes Written=47
[hadoop@localhost hadoop]$
```

图5-14　WordCount.jar运行效果

项目5　MapReduce 编程

> **扩展**
>
> 尝试一下全分布模式下的运行效果？
>
> 1）在伪分布式节点上使用 scp 指令将 jar 包发送到全分布式环境中的 master 节点上，如图 5-13 所示。
>
> 　$ cd /usr/local/hadoop
> 　$ scp WordCount.jar hadoop@master节点IP:/usr/local/hadoop
>
> ```
> [root@localhost hadoop]# scp WordCount.jar hadoop@192.1
> 68.247.194:/usr/local/hadoop
> hadoop@192.168.247.194's password:
> WordCount.jar        100% 3304     2.5MB/s   00:00
> [root@localhost hadoop]#
> ```
>
> 图5-13　传输WordCount.jar到全分布式环境
>
> 2）在 master 节点使用 start-all.sh 启动全分布式下所有 Hadoop 进程。
>
> 3）在 master 节点上使用同样的方式在 HDFS 上创建 input 文件夹及两个任务文件 wordfile1.txt 与 wordfile2.txt。
>
> 4）在 master 节点上使用相同指令执行 WordCount.jar。执行效果如图 5-15 所示。
>
> ```
> 2022-10-05 22:16:32,515 INFO mapreduce.Job: Counters: 54
>     File System Counters
>         FILE: Number of bytes read=106
>         FILE: Number of bytes written=828190
>         FILE: Number of read operations=0
>         FILE: Number of large read operations=0
>         FILE: Number of write operations=0
>         HDFS: Number of bytes read=286
>         HDFS: Number of bytes written=47
>         HDFS: Number of read operations=11
>         HDFS: Number of large read operations=0
>         HDFS: Number of write operations=2
>         HDFS: Number of bytes read erasure-coded=0
>     Job Counters
>         Launched map tasks=2
>         Launched reduce tasks=1
>         Data-local map tasks=2
>         Total time spent by all maps in occupied slots (ms)=8323
>         Total time spent by all reduces in occupied slots (ms)=2202
>         Total time spent by all map tasks (ms)=8323
>         Total time spent by all reduce tasks (ms)=2202
>         Total vcore-milliseconds taken by all map tasks=8323
>         Total vcore-milliseconds taken by all reduce tasks=2202
>         Total megabyte-milliseconds taken by all map tasks=8522752
>         Total megabyte-milliseconds taken by all reduce tasks=2254848
>     Map-Reduce Framework
>         Map input records=4
>         Map output records=12
>         Map output bytes=104
>         Map output materialized bytes=112
>         Input split bytes=230
>         Combine input records=12
>         Combine output records=9
>         Reduce input groups=7
>         Reduce shuffle bytes=112
>         Reduce input records=9
>         Reduce output records=7
>         Spilled Records=18
>         Shuffled Maps =0
>         Failed Shuffles=0
>         Merged Map outputs=2
>         GC time elapsed (ms)=263
>         CPU time spent (ms)=1160
>         Physical memory (bytes) snapshot=873025536
>         Virtual memory (bytes) snapshot=8938237952
>         Total committed heap usage (bytes)=709361664
>         Peak Map Physical memory (bytes)=330448896
>         Peak Map Virtual memory (bytes)=2977320960
>         Peak Reduce Physical memory (bytes)=223068160
>         Peak Reduce Virtual memory (bytes)=2984706048
>     Shuffle Errors
>         BAD_ID=0
>         CONNECTION=0
>         IO_ERROR=0
>         WRONG_LENGTH=0
>         WRONG_MAP=0
>         WRONG_REDUCE=0
>     File Input Format Counters
>         Bytes Read=56
>     File Output Format Counters
>         Bytes Written=47
> ```
>
> 图5-15　WordCount.jar全分布模式运行效果
>
> 可以对比看看两者运行时输出内容的差异。

> **注意**
>
> 如果要再次运行 WordCount.jar，需要先删除 HDFS 中的 output 目录，否则会报错。

上面命令执行以后，当运行顺利结束时，词频统计结果已经被写入了 HDFS 的 "/user/hadoop/output" 目录中，执行如下命令查看词频统计结果，如图5-16所示。

```
$ hdfs dfs -cat output/*
```

```
[hadoop@localhost hadoop]$ hdfs dfs -cat output/*
Hadoop  2
I       2
Spark   2
fast    1
good    1
is      2
love    2
```

图5-16　WordCount.jar输出结果

拓展学习

1. MapReduce工作流程

在MapReduce计算框架下，开发人员仅需通过编写map()方法与reduce()方法即可实现相应的批处理逻辑，为大数据的并行计算提供便利。那么，MapReduce计算框架是如何开展工作的呢？MapReduce运行机制如图5-17所示。

扫码看视频

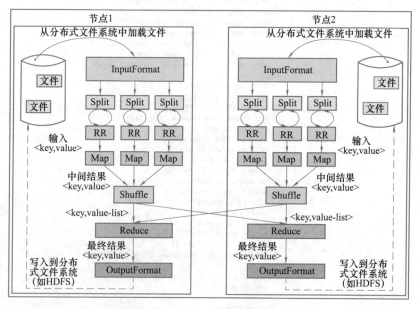

图5-17　MapReduce运行机制

（1）数据预处理

MapReduce计算框架首先加载分布式文件系统（如HDFS）中的文件，并使用InputFormat模块实现数据预处理，包括：

1）验证输入的格式是否符合输入定义。

2）将源文件在逻辑上划分为小数据块分片（Split）。逻辑划分，即没有真正将文件进行切分，而只是记录了数据块的位置与长度信息。Split的数量即为后续Map阶段的Map任务数量。

由于Split分片是基于尺寸的分割，因此需要借助RecordReader（RR），基于Split的内容信息，将数据加载并转换为适合Map任务读取的键值对<key,value>形式。输入给下一步的Map任务。

（2）Map任务

Map任务会根据用户自定义的映射规则，输出一系列的<key,value>作为中间结果。中间结果以文件形式存储于本地存储设备（如磁盘）。

（3）Shuffle过程

将Map任务得到的中间结果数据，按key进行分区、排序、合并、归并等操作，使数据形式从List(<key,value>)转变为<key,value-list>，最后分发给相应的Reduce任务。

（4）Reduce任务

Reduce任务以<key,value-list>作为输入，根据用户自定义的Reduce规则进行逻辑处理，输出<key,value>的最终结果，交给OutputFormat模块。

（5）输出结果

OutputFormat模块会验证输出目录是否存在、是否符合配置文件要求。若全都满足，则输出Reduce的结果到分布式文件系统。

Shuffle过程的性能高低直接决定了整个MapReduce程序的性能高低。Shuffle过程分为Map端Shuffle和Reduce端Shuffle，如图5-18所示，具体如下：

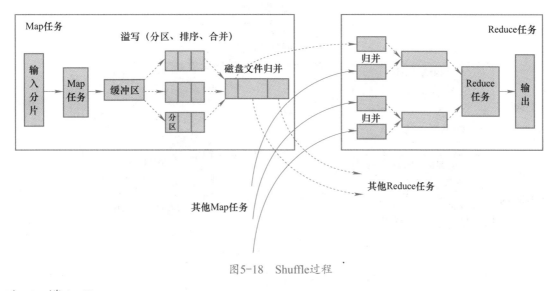

图5-18　Shuffle过程

1）Map端shuffle：

① 溢写：每一个Map任务都有一个内存缓冲区（默认为100MB），用于存放Map任务输出的中间结果。如果写入的数据达到内存缓冲区设定的阈值（默认为80%，即80MB）时，会启动溢写操作，将内存数据写入磁盘。

② 分区：写入磁盘之前，按照Reduce任务的个数对缓冲区数据进行分区。分区的目的是避免一些Reduce任务分配的数据过多，而其他Reduce任务分配的数据过少。

③ 排序、合并：分区后，对每个分区的数据进行排序和合并。合并的目的是尽量减少写入磁盘的数据量。

④ 归并：如果中间结果数据较多，则会生成多个溢写文件。最后的缓冲区数据也会全部溢写。所有溢写文件会在最后归并成一个单独文件。归并的过程也需要进行排序、合并操作，目的是减少数据量，既能减少磁盘IO时间，又能降低网络传输延迟。

⑤ 通知相应的Reduce任务来"领取"分配给自己的数据。

2）Reduce端shuffle：

① 领取数据：Reduce任务会通过RPC向JobTracker询问各个不同的Map任务是否完成，若完成，则领取数据。

② 溢写：领取的数据存入相应缓冲区，如果超出缓冲区的阈值，则对数据进行归并、溢写。Map任务的输出数据已经是排好序的，当开启溢写操作，相同key的键值对会被归并，在磁盘中产生一个一个溢写文件。

③ 归并：随着Map数据全部被领取，多个溢写产生的多个溢写文件会被归并成一个大文件，归并仍会对键值对进行排序。

④ 输出：最后一次归并的数据，直接输入到Reduce任务。

2. 词频统计实例执行过程

WordCount执行过程如图5-19所示。

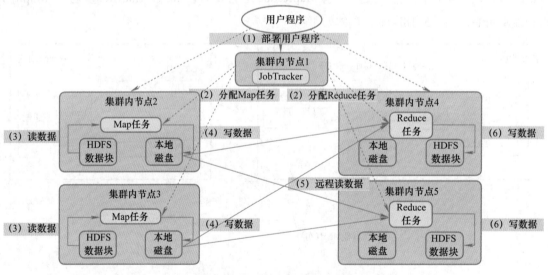

图5-19　WordCount执行过程

1）部署用户程序：系统将用户程序部署在集群中的多个节点主机中，其中，运行JobTracker的为主节点，负责调度作业，其余节点为从节点，负责执行Map任务与Reduce任务。

2）分配Map/Reduce任务：按照"计算向数据靠拢"的设计理念，系统会按照文件数据块所在节点为用户程序分配N个执行Map任务的节点，另外，根据系统资源使用情况，分配M个执行Reduce任务的节点。

3）读数据：Map任务读取HDFS数据块，并执行Map逻辑，输出的中间结果写在内存缓冲区中。

4）写数据（中间结果）：缓冲区的中间结果会定期写入本地磁盘，并被划分为M个分区，准备后续分发给M个Reduce任务。JobTracker会记录每个Map任务的中间结果的每个分区位置，并告知Reduce任务领取。

5）远程读数据：Reduce任务从每个Map任务的本地磁盘中领取中间结果，并对所有结果进行排序，使相同的key的键值对聚集在一起。

6）写数据（最终结果）：针对每一个唯一的key，执行Reduce逻辑，结果写入到输出文件，输出文件存储于HDFS文件系统中。

7）返回结果：唤醒用户程序，返回结果。

3. 常用Java API介绍

MapReduce编程时，可以在官方文档（https://hadoop.apache.org/docs/stable/api/index.html）中查看并学习API，其中，常用的Java API见表5-3。

表5-3 常用API介绍

类/接口名称	描 述
InputFormat	描述了MapReduce作业的输入规范 MapReduce框架依靠作业的InputFormat实现： 1）验证作业的输入规范； 2）将输入文件分割成逻辑的InputSplits； 3）将每个文件分配给各个Mapper
OutputFormat	描述了MapReduce作业的输出规范 MapReduce框架依靠作业的OutputFormat实现： 1）验证作业的输出规范；例如，检查输出目录是否已经存在； 2）生成作业的输出文件，并将输出文件存储在一个文件系统中
Writable	基于DataInput和DataOutput实现了一个简单、高效的序列化协议。Hadoop MapReduce框架中的任何键或值类型都实现了这个接口，通常包括数据的读写需求
WritableComparable	在Writable基础上继承了Comparable接口，通过比较器实现相互比较。由于key包含了比较排序的操作，因此，任何要在Hadoop MapReduce框架中作为key使用的类型都应该实现这个接口

更多Java API信息请参考官网，大家要培养从官网查询并学习技术文档的能力。

项目小结

本项目针对MapReduce编程的典型工作任务，学习了MapReduce核心思想、设计理念、体系结构、编程思想，了解了Hadoop的版本变迁中，MapReduce与Yarn的关系，并编写程序实现了词频统计功能。完成任务后，进一步了解了与MapReduce相关的知识，加深对MapReduce的理解。

实战强化

某同学在学习MapReduce经典程序WordCount时，执行WordCount.jar却得到了如图5-20所示内容。

```
[hadoop@localhost hadoop]$ hadoop jar ./WordCount.jar WordCount input output
2022-10-05 22:59:04,282 INFO impl.MetricsConfig: Loaded properties from hadoop-metr
ics2.properties
2022-10-05 22:59:04,389 INFO impl.MetricsSystemImpl: Scheduled Metric snapshot peri
od at 10 second(s).
2022-10-05 22:59:04,389 INFO impl.MetricsSystemImpl: JobTracker metrics system star
ted
Exception in thread "main" org.apache.hadoop.mapred.FileAlreadyExistsException: Out
put directory hdfs://localhost:9000/user/hadoop/output already exists
        at org.apache.hadoop.mapreduce.lib.output.FileOutputFormat.checkOutputSpecs
(FileOutputFormat.java:164)
        at org.apache.hadoop.mapreduce.JobSubmitter.checkSpecs(JobSubmitter.java:27
7)
        at org.apache.hadoop.mapreduce.JobSubmitter.submitJobInternal(JobSubmitter.
java:143)
        at org.apache.hadoop.mapreduce.Job$11.run(Job.java:1571)
        at org.apache.hadoop.mapreduce.Job$11.run(Job.java:1568)
        at java.security.AccessController.doPrivileged(Native Method)
        at javax.security.auth.Subject.doAs(Subject.java:422)
        at org.apache.hadoop.security.UserGroupInformation.doAs(UserGroupInformatio
n.java:1878)
```

图5-20　WordCount执行过程出错

试分析系统报错的原因，并给出解决方案。

项目 6　部署与使用 HBase

项目概述

A公司根据客户的大数据业务场景需求，开发完成了一套基于Hadoop的大数据平台项目。客户需要在大数据平台中构建分布式数据表，用于存储大量业务数据，A公司拟为其部署HBase分布式数据库以实现客户需求。公司委派大数据运维工程师王工前往客户的中心机房对HBase进行集群部署。

本项目针对"部署与使用HBase"的典型工作任务，完成三个工作任务：部署伪分布式HBase、部署完全分布式HBase、利用Shell命令操作HBase。

在开展任务前，需要掌握必要的理论知识：什么是HBase？HBase与其他组件的关系是怎样的？数据模型是怎样的？体系架构是怎样的？如何使用Shell命令进行HBase数据操作？

完成任务后，深入了解HBase内部工作机制，包括Region服务器概念、HLog与Store原理。

学习目标

1. 熟悉并理解HBase的特点。
2. 掌握HBase的主要组成结构。
3. 了解HBase的实现原理。
4. 熟练掌握伪分布式HBase部署。
5. 基本掌握完全分布式HBase部署。
6. 熟练使用HBase常用的Shell命令。

思维导图

项目思维导图如图6-1所示。

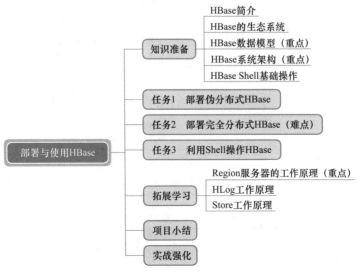

图6-1　项目思维导图

知识准备

1. HBase简介

扫码看视频

分布式存储最早源于谷歌的互联网搜索引擎业务，将爬虫不断爬取的新页面每页一行地存储到BigTable里。BigTable是谷歌公司内部使用的分布式存储数据库。HBase是一个高可靠、高性能、面向列、可伸缩的分布式数据库，是谷歌BigTable的开源实现，主要用来存储非结构化和半结构化的松散数据。HBase的目标是处理非常庞大的表，可以通过水平扩展的方式，利用廉价计算机集群处理由超过10亿行数据和数百万列元素组成的数据表。

（1）HBase和BigTable的关系

HBase是BigTable的开源实现，它们的底层技术对应关系见表6-1。

表6-1　HBase和BigTable底层技术关系

项　目	BigTable	HBase
文件存储系统	GFS	HDFS
海量数据处理	MapReduce	Hadoop MapReduce
协同管理服务	Chubby	ZooKeeper

（2）HBase项目意义

关系数据库已经流行很多年，并且Hadoop已经有了HDFS和MapReduce，为什么需要HBase呢？原因如下：

1）Hadoop可以很好地解决大规模数据的离线批量处理问题，但是受限于Hadoop MapReduce编程框架的高延迟数据处理机制，Hadoop无法满足大规模数据实时处理应用的需求。

2）HDFS面向批量访问模式，不是随机访问模式。

3）传统的通用关系型数据库无法应对在数据规模剧增时导致的系统扩展性和性能问题，即使是分库分表也不能很好解决。

2. HBase的生态系统

HBase与Hadoop生态系统中的其他组件的关系如图6-2所示。

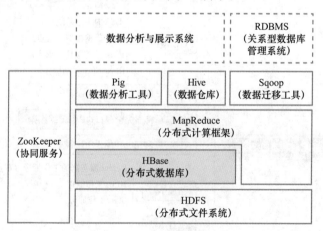

图6-2　HBase与Hadoop生态系统

图6-2中，HBase与各部分组件的关系见表6-2。

项目6 部署与使用HBase

表6-2 HBase与Hadoop生态组件关系

组 件	关 系 描 述
HDFS（分布式文件系统）	相较于将数据存储于本地文件系统，HBase可以将海量数据存储于HDFS文件系统中，实现高可靠存储，更好地支持大量数据的处理
MapReduce（分布式计算框架）	为HBase提供计算能力，处理HBase中的海量数据，实现高性能计算
ZooKeeper（协同服务）	为HBase提供稳定服务以及任务的失败恢复
Sqoop（数据迁移工具）	高效、便捷地将传统关系型数据库管理系统中的数据迁移至HBase分布式数据库中
Pig（数据分析工具）	为HBase提供高级语言支持。使用者无需编写复杂的MapReduce代码，仅需使用Pig Latin或HiveQL这样简单的脚本语句，即可实现对HBase数据库中数据的管理、查询，为上层数据分析与展示系统提供数据支持
Hive（数据仓库）	

3. HBase数据模型

HBase主要用于存储非结构化和半结构化的数据。认识HBase数据模型要从以下几个方面着手：

1）HBase是一个稀疏、多维度、排序的映射表，这张表的索引是行键、列族、列限定符和时间戳。

2）每个值是一个未经解释的字符串，没有数据类型。

3）列族支持动态扩展，可以很轻松地添加一个列族或列，无须预先定义。

4）HBase中执行更新操作时，并不会删除数据旧的版本，而是生成一个新的版本，旧版本仍然保留（这是和HDFS只允许追加不允许修改的特性相关的）。

HBase每个表由行和列组成，列划分为若干个列族，列族里的数据通过列限定符（或列）来定位；每个行由行键（row key）来标识。在HBase表中，通过行键、列族和列限定符确定一个"单元格"，单元格中存储的数据没有数据类型，总被视为字节数组byte[]；每个单元格都保存着同一份数据的多个版本，这些版本采用时间戳进行索引辨别。

由此，确定和引用某个单元格在某一时间点上的数据，可以用四维坐标去描述，分别是：行键、列族、列限定符和时间戳，如图6-3所示。

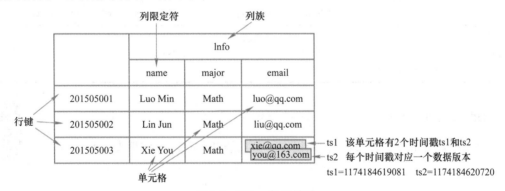

图6-3 HBase数据模型

在上述学生信息表中，如果要定位Xie You学生的最新E-mail值，需要使用四维坐标["201505003", "Info", "email", "1174184620720"]来确定表内对应位置的值为：you@163.com。

4. HBase系统架构

HBase系统架构如图6-4所示，架构在底层的Hadoop集群之上，采用与Hadoop集群相同的主从架构，数据存储并不直接和底层的磁盘打交道，而是借助于HDFS完成数据存储，整个系统架构的主要组件：ZooKeeper服务器、Master服务器（HMaster）、Region服务器（HRegionServer）和客户端。

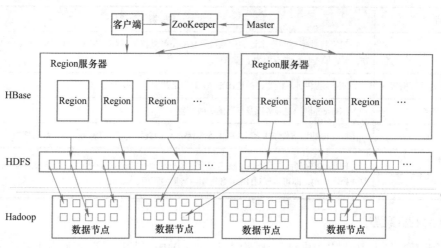

图6-4 HBase系统架构图

各组件的作用见表6-3。

表6-3 HBase各组件作用

组 件	作 用
HMaster	集群中会有多个HMaster，但同一时间仅有一个处于工作状态，其余处于待命状态。 HMaster充当整个HBase集群中的管家角色： 1）管理HRegionServer列表； 2）为空闲的HRegionServer分配数据，实现负载均衡； 3）为故障的HRegionServer上的数据重新分配新的HRegionServer，并处理HRegionServer上的故障问题； 4）管理用户对表的增加、删除、修改、查询等操作
ZooKeeper	ZooKeeper充当整个HBase集群中的管家助理角色，帮助管家实现对集群的协同服务： 1）HRegionServer以心跳形式告知ZooKeeper自身状态，ZooKeeper将异常情况通知HMaster； 2）从多个HMaster中选择唯一一个HMaster作为工作服务器，其余为待命服务器； 3）保存用户数据表访问地址和HMaster主服务器的地址，实现客户端访问数据的同时降低HMaster负载
HRegionServer	HRegionServer是HBase集群中的核心模块，功能如下： 1）维护分配给自己的数据，将数据实际存储于HDFS内，由HDFS实现高可靠存储； 2）响应用户的读写请求

5. HBase Shell基础操作

HBase支持通过Shell的方式实现数据库操作，常见的Shell基础操作见表6-4。

扫码看视频

表6-4 Shell基础操作

操作类型	Shell命令	功 能
数据定义语言（操作数据表）	create	创建表
	d<esc>ribe	显示表的相关信息
	list	列出HBase中所有表信息
	enable/disable	使表有效/无效
	drop	删除表

（续）

操作类型	Shell命令	功 能
数据操作语言（操作数据）	put	向表内指定单元格添加数据
	get	根据指定表名、行键、列族、列限定符、时间戳、版本号获得单元格的值
	scan	浏览表相关信息
	count	统计表中行数
	delete	删除表中指定单元格的数据
	truncate	删除表，并重新建立表

任务1 部署伪分布式HBase

扫码看视频

任务描述

王工在前往客户现场前，为了确保顺利实现HBase的集群部署，先在自己公司的单机环境下进行伪分布式部署，并测试HBase的基础功能。

任务分析

1. 任务目标

能在已完成的伪分布式Hadoop中部署HBase伪分布式模式。

2. 任务环境

操作系统：CentOS Stream 9（预装伪分布式Hadoop）

软件版本：Java 1.8.0、Hadoop 3.3.4、HBase 2.4.14

3. 任务导图

任务导图如图6-5所示。

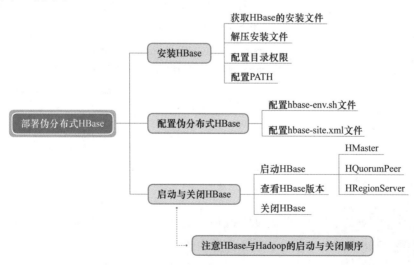

图6-5 任务导图

任务实施

1. 安装HBase

（1）获取HBase的安装文件

到Apache官网提供的下载地址（https://dlcdn.apache.org）找到hbase，选择要下载的版本，本书使用的版本为2.4.14。

```
# cd /usr/local
#wget https://dlcdn.apache.org/hbase/2.4.14/hbase-2.4.14-bin.tar.gz
```

（2）解压安装文件

```
#tar -zxf hbase-2.4.14-bin.tar.gz
```

更改目录名称为hbase，如图6-6所示。

```
#mv hbase-2.4.14 hbase
```

```
[root@localhost local]# ls
bin   games   hadoop-3.3.4.tar.gz           include   lib64     sbin    src
etc   hadoop  hbase-2.4.14-bin.tar.gz       lib       libexec   share
[root@localhost local]# tar -zxf hbase-2.4.14-bin.tar.gz -C /usr/local
[root@localhost local]# ls
bin   games   hadoop-3.3.4.tar.gz   hbase-2.4.14-bin.tar.gz   lib       libexec   share
etc   hadoop  hbase-2.4.14          include                   lib64     sbin      src
[root@localhost local]# mv hbase-2.4.14 hbase
[root@localhost local]# ls
bin   games   hadoop-3.3.4.tar.gz   hbase-2.4.14-bin.tar.gz   lib       libexec   share
etc   hadoop  hbase                 include                   lib64     sbin      src
```

图6-6 解压并重命名hbase

（3）配置目录权限

将目录的所有者改为hadoop用户，如图6-7所示。

```
# chown -R hadoop:hadoop hbase
```

```
[root@localhost local]# chown -R hadoop:hadoop hbase
[root@localhost local]# ls -ld hbase
drwxr-xr-x. 7 hadoop hadoop 182 10月  6 01:27 hbase
```

图6-7 更改目录权限

（4）配置PATH

使用vim编辑器编辑/etc/profile。

```
# vim /etc/profile
```

在打开的profile文件中，添加HBase相关路径。在末行增加如下内容，如图6-8所示。

```
export HBASE_HOME=/usr/local/hbase
export PATH=$PATH:$HBASE_HOME/bin
```

```
export JAVA_HOME=/usr/lib/jvm/java-1.8.0-openjdk-1.8.
0.345.b01-5.el9.x86_64/jre
export PATH=$PATH:$JAVA_HOME

export HADOOP_HOME=/usr/local/hadoop
export PATH=$PATH:$HADOOP_HOME/bin:$HADOOP_HOME/sbin

export HBASE_HOME=/usr/local/hbase
export PATH=$PATH:$HBASE_HOME/bin
```

图6-8 配置环境变量

配置完成后，按<ESC>键，输入:wq，保存并退出。使用source指令，使新配置生效。

```
# source /etc/profile
```

说明

随着版本更新，当前版本号的下载地址可能发生变化，请读者自行到官网获取相应下载地址。

说明

如果通过wget指令获取安装包太慢，可以使用本书提供的安装包。使用scp指令，从Windows传输文件至Linux中。

笔记

HBase有哪几种部署形式？

HBase与Hadoop相同，有三种部署形式：单机模式、伪分布模式和完全分布模式。

注意

更新/etc/profile后务必使用source指令使更新后的内容生效。

2. 配置伪分布式HBase

切换为hadoop用户。

su - hadoop

（1）配置hbase-env.sh文件

编辑/usr/local/hbase/conf/hbase-env.sh。

$ vim /usr/local/hbase/conf/hbase-env.sh

1）将export JAVA_HOME前面的#去掉，使变量生效，并为其指定正确的JAVA_HOME路径，如图6-9所示。

```
# The java implementation to use.  Java 1.8+ required.
export JAVA_HOME=/usr/lib/jvm/java-1.8.0-openjdk-1.8.0.345.b01-2.el9.x86_64/jre
```

图6-9　设置JAVA_HOME

> ⚠️ 注意
> 本任务中的JAVA_HOME路径仅作参考，请使用自己实际的JAVA_HOME路径。

2）去掉export HBASE_MANAGES_ZK=true前面的#，使其生效，让HBase使用自带的ZooKeeper，如图6-10所示。

```
# Tell HBase whether it should m
export HBASE_MANAGES_ZK=true

# The default log rolling policy
```

图6-10　设置HBASE_MANAGES_ZK

> 📝 笔记
> HBase的运行必须依赖ZooKeeper。配置项HBASE_MANAGES_ZK值为true表示使用自带ZooKeeper，false表示使用集群另外配置的ZooKeeper组件。

修改完成后，保存并退出。

（2）配置hbase-site.xml文件

打开hbase-site.xml配置文件。

$ vim /usr/local/hbase/conf/hbase-site.xml

配置如下内容，如图6-11所示。

```xml
<property>
    <name>hbase.cluster.distributed</name>
    <value>true</value>
</property>
<property>
    <name>hbase.tmp.dir</name>
    <value>/usr/local/hbase/data</value>
</property>
<property>
    <name>hbase.unsafe.stream.capability.enforce</name>
    <value>false</value>
</property>
<property>
    <name>hbase.rootdir</name>
    <value>hdfs://localhost:9000/hbase</value>
</property>
<property>
    <name>hbase.zookeeper.quorum</name>
    <value>localhost</value>
</property>
```

> 📝 笔记
> hbase.cluster.distributed，用于设置HBase是否为分布式。

> 📝 笔记
> hbase.tmp.dir，用于设置HBase临时文件路径。

> 📝 笔记
> hbase.rootdir，用于设置HBase中用户数据对应的文件存放的位置。本书设置为存放在HDFS分布式文件系统根路径下的hbase目录中。

> 📝 笔记
> hbase.zookeeper.quorum，用于设置ZooKeeper节点列表。

```xml
<property>
    <name>hbase.cluster.distributed</name>
    <value>true</value>
</property>
<property>
    <name>hbase.tmp.dir</name>
    <value>/usr/local/hbase/data</value>
</property>
<property>
    <name>hbase.unsafe.stream.capability.enforce</name>
    <value>false</value>
</property>
<property>
    <name>hbase.rootdir</name>
    <value>hdfs://localhost:9000/hbase</value>
</property>
<property>
    <name>hbase.zookeeper.quorum</name>
    <value>localhost</value>
</property>
```

图6-11　设置伪分布式hbase-site.xml

3. 启动与关闭HBase

（1）启动HBase

启动HBase前，先启动Hadoop，并看到NameNode、DataNode、SecondaryNameNode进程。

```
$ start-dfs.sh
```

然后启动HBase，启动成功后如图6-12所示，有HMaster、HQuorumPeer、HRegionServer三个进程。

```
$ start-hbase.sh
```

```
[hadoop@localhost ~]$ jps
19187 Jps
16420 DataNode
18692 HMaster
16263 NameNode
18569 HQuorumPeer
16682 SecondaryNameNode
18861 HRegionServer
```

图6-12　启动HBase

> ⚠ 注意
>
> 由于HBase需要使用HDFS分布式文件系统，因此，可以认为HBase是部署在Hadoop集群之上的。需要保证在HBase运行的整个生命周期中，Hadoop都是正常运行的。所以，启动与关闭HBase的顺序是：启动Hadoop→启动HBase→关闭HBase→关闭Hadoop。

> 📝 笔记
>
> 进程中，HMaster对应Master服务器；HQuorumPeer对应HBase自带的ZooKeeper；HRegionServer对应Region服务器。伪分布模式下，三者部署在一个节点上。

（2）查看HBase版本

查看HBase版本，如图6-13所示。

```
$ hbase version
```

```
[hadoop@localhost conf]$ hbase version
HBase 2.4.14
Source code repository git://7fae9230bab5/home/hsun/hbase-rm/output/hbase revision=2e7d75a892000071a7479b2f668c4db7a241be3f
Compiled by hsun on Tue Aug 23 23:33:09 UTC 2022
From source with checksum 70fed5374592514310c413f7af4817dd2e3947ef2315b9ca317fc2ebfe0a63bb114239d5adda5db289748669574cadc849007caf52a4098420e16a50b7f21242
```

图6-13　查看HBase版本

（3）关闭HBase

关闭HBase，如图6-14所示。

```
$ stop-hbase.sh
```

```
[hadoop@localhost hbase]$ stop-hbase.sh
stopping hbase.........
localhost: running zookeeper, logging to /usr/local/hbase/bin/../logs/hbase-hadoop-zookeeper-localhost.localdomain.out
localhost: stopping zookeeper.
[hadoop@localhost hbase]$ jps
151539 DataNode
151796 SecondaryNameNode
153461 Jps
151368 NameNode
```

图6-14　关闭HBase

> ⚠ 注意
>
> 运行HBase的过程中，如果提示"SLF4J"，这是由于HBase和Hadoop的jar包发生冲突，可以在hbase-env.sh配置文件中去掉export HBASE_DISABLE_HADOOP_CLASSPATH_LOOKUP="true"前的#，使其生效，如图6-15所示。重启HBase即可。

```
# the default value is false,means that includes Had
export HBASE_DISABLE_HADOOP_CLASSPATH_LOOKUP="true"
```

图6-15　更新配置信息

项目6 部署与使用HBase

扫码看视频

任务2　部署完全分布式HBase

任务描述

王工来到客户现场，现在客户的集群环境中已经有完全分布式Hadoop了，接下来他要为客户部署完全分布式HBase。

任务分析

1. 任务目标

能在已完成的完全分布式Hadoop中部署完全分布式HBase。

2. 任务环境

操作系统：CentOS Stream 9（预装完全分布式Hadoop）

软件版本：Java 1.8.0、Hadoop 3.3.4、HBase 2.4.14

3. 任务导图

任务导图如图6-16所示。

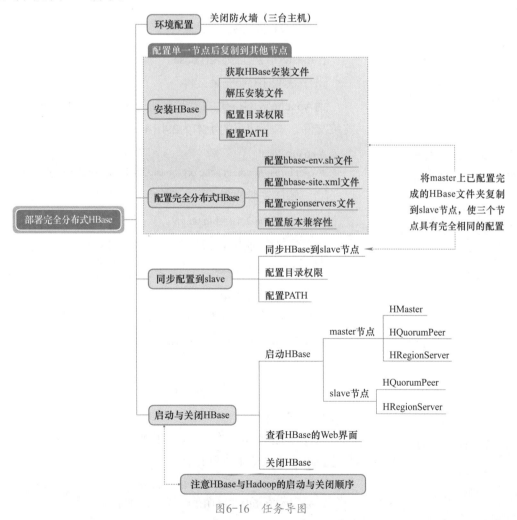

图6-16　任务导图

任务实施

1. 环境配置

关闭防火墙（三台主机）

运行指令，关闭三个节点的防火墙，并设置开机不自动启动防火墙，查看防火墙状态，确认防火墙已关闭，如图6-17所示。

```
# systemctl stop firewalld
# systemctl disable firewalld
# systemctl status firewalld
```

图6-17 关闭防火墙

📝 **笔记**

为什么关闭防火墙？

在完全分布式环境下，ZooKeeper服务需要访问各节点的2181端口，关闭防火墙可以保证各节点的端口能正常访问。

2. 安装HBase

在master节点上安装HBase。

（1）获取HBase安装文件

到Apache官网提供的下载地址（https://dlcdn.apache.org）下载安装文件，本书使用的版本为2.4.14。

```
# cd /usr/local
#wget https://dlcdn.apache.org/hbase/2.4.14/hbase-2.4.14-bin.tar.gz
```

（2）解压安装文件

```
#tar -zxf hbase-2.4.14-bin.tar.gz
```

更改目录名称为hbase。

```
#mv hbase-2.4.14 hbase
```

（3）配置目录权限

将目录的所有者改为hadoop用户。

```
# chown -R hadoop:hadoop hbase
```

（4）配置PATH

使用vim编辑器编辑/etc/profile。

```
# vim /etc/profile
```

在打开的profile文件中，添加HBase相关路径。

在末行增加如下内容：

```
export HBASE_HOME=/usr/local/hbase
export PATH=$PATH:$HBASE_HOME/bin
```

📝 **说明**

安装HBase过程与伪分布式相同，详细内容参照任务1，仅罗列必要步骤。

📝 **说明**

如果通过wget指令获取安装包太慢，可以使用本书提供的安装包。使用scp指令，从Windows传输文件至Linux中。

配置完成后，按<ESC>键，输入:wq，保存并退出。使用source指令，使新配置生效。

source /etc/profile

3. 配置完全分布式HBase

在master节点，切换为hadoop用户。

su - hadoop

（1）配置hbase-env.sh文件

编辑/usr/local/hbase/conf/hbase-env.sh。

$ vim /usr/local/hbase/conf/hbase-env.sh

1）将export JAVA_HOME前面的#去掉，并配置实际JAVA_HOME路径。

2）去掉export HBASE_MANAGES_ZK=true前面的#，使其生效，让HBase使用自带的ZooKeeper。

3）去掉export HBASE_DISABLE_HADOOP_CLASSPATH_LOOKUP="true"前面的#号，使其生效，使HBase在启动时不包括Hadoop的lib包，避免包冲突，如图6-18所示。

```
# the default value is false,means that includes Had
export HBASE_DISABLE_HADOOP_CLASSPATH_LOOKUP="true"
```

图6-18 设置hbase-env.sh配置信息

修改完成后，保存并退出。

（2）配置hbase-site.xml文件

打开/usr/local/hbase/conf/hbase-site.xml配置文件，如图6-19所示。

$ vim /usr/local/hbase/conf/hbase-site.xml

配置如下内容：

```xml
<property>
    <name>hbase.cluster.distributed</name>
    <value>true</value>
</property>
<property>
    <name>hbase.tmp.dir</name>
    <value>/usr/local/hbase/data</value>
</property>
<property>
    <name>hbase.unsafe.stream.capability.enforce</name>
    <value>false</value>
</property>
<property>
    <name>hbase.rootdir</name>
    <value>hdfs://master:9000/hbase</value>
</property>
<property>
```

⚠️ **注意**

本任务中的JAVA_HOME路径仅作参考，请使用自己实际的JAVA_HOME路径。

📝 **笔记**

HBase的运行必须依赖ZooKeeper。本任务采用HBase自带的ZooKeeper。但是，在实际生产环境下，HBase更多会搭配独立的ZooKeeper组件，提高平台可用性。感兴趣的读者可以自行下载并配置ZooKeeper。

📝 **笔记**

hbase.cluster.distributed，用于设置HBase是否为分布式。

📝 **笔记**

hbase.tmp.dir，用于设置HBase在本地生成的临时文件存放的路径。

📝 **笔记**

hbase.rootdir，用于设置HBase中用户数据对应的文件存放的位置，此处设置为存放在HDFS分布式文件系统根路径下的hbase目录中。

笔记

hbase.zookeeper.quorum，用于设置zookeeper节点列表。

 笔记

hbase.master，用于设置主节点的访问地址。

笔记

hbase.zookeeper.property.dataDir，用于表示HBase在ZooKeeper上存放数据的位置。

```xml
<name>hbase.zookeeper.quorum</name>
<value>master:2181,slave1:2181,slave2:2181</value>
</property>
<property>
    <name>hbase.master </name>
    <value>hdfs://master:60000</value>
</property>
<property>
    <name>hbase.zookeeper.property.dataDir </name>
    <value>/usr/local/hbase/zoodata</value>
</property>
```

```xml
<property>
  <name>hbase.cluster.distributed</name>
  <value>true</value>
</property>
<property>
  <name>hbase.tmp.dir</name>
  <value>/usr/local/hbase/data</value>
</property>
<property>
  <name>hbase.unsafe.stream.capability.enforce</name>
  <value>false</value>
</property>
<property>
  <name>hbase.rootdir</name>
  <value>hdfs://master:9000/hbase</value>
</property>
<property>
  <name>hbase.zookeeper.quorum</name>
  <value>master:2181,slave1:2181,slave2:2181</value>
</property>
<property>
  <name>hbase.master</name>
  <value>hdfs://master:60000</value>
</property>
<property>
  <name>hbase.zookeeper.property.dataDir</name>
  <value>/usr/local/hbase/zoodata</value>
</property>
```

图6-19 完全分布式配置信息

（3）配置regionservers文件

修改regionservers文件。

```
$ vi regionservers
```

添加如下内容：

```
master
slave1
slave2
```

 笔记

regionservers配置文件用于展示系统中所有的Region服务器的IP地址列表。

此处，用主机名代替IP地址，有利于集群的长期维护。

（4）配置版本兼容性

由于本书选择的HBase 2.4.14版本与Hadoop 3.3.4版本存在兼容性问题，此处使用Hadoop的包替换HBase的包。

先对原HBase的jar包进行备份。

```
$ cd /usr/local/hbase/lib/client-facing-thirdparty
$ mkdir bak
$ mv slf4j* bak
```

说明

由于每个版本具有各自的特殊性，各版本的兼容性配置处理方式不同。

复制hadoop路径下的对应jar包至hbase路径下：

```
$ cp /usr/local/hadoop/share/hadoop/common/lib/slf4j-* .
```

如图6-20所示，slf4j的jar包版本从1.7.33替换成了1.7.36。

```
[hadoop@master conf]$ cd /usr/local/hbase/lib/client-facing-thirdp
arty/
[hadoop@master client-facing-thirdparty]$ ls
audience-annotations-0.5.0.jar    reload4j-1.2.22.jar
commons-logging-1.2.jar           slf4j-api-1.7.33.jar
htrace-core4-4.2.0-incubating.jar slf4j-reload4j-1.7.33.jar
[hadoop@master client-facing-thirdparty]$ mkdir bak
[hadoop@master client-facing-thirdparty]$ mv slf4j-* bak
[hadoop@master client-facing-thirdparty]$ ls bak/
slf4j-api-1.7.33.jar  slf4j-reload4j-1.7.33.jar
[hadoop@master client-facing-thirdparty]$ ls
audience-annotations-0.5.0.jar    htrace-core4-4.2.0-incubating.jar
bak                               reload4j-1.2.22.jar
commons-logging-1.2.jar
[hadoop@master client-facing-thirdparty]$ cp /usr/local/hadoop/sha
re/hadoop/common/lib/slf4j-* .
[hadoop@master client-facing-thirdparty]$ ls
audience-annotations-0.5.0.jar    reload4j-1.2.22.jar
bak                               slf4j-api-1.7.36.jar
commons-logging-1.2.jar           slf4j-reload4j-1.7.36.jar
htrace-core4-4.2.0-incubating.jar
```

图6-20　兼容性配置

4．同步配置到slave

（1）同步HBase到slave节点

分别复制hbase文件夹至slave1、slave2节点。如图6-21所示为复制至slave1节点、slave2节点的过程与之类似。

$ scp -r /usr/local/hbase root@slave1:/usr/local
$ scp -r /usr/local/hbase root@slave2:/usr/local

```
[hadoop@master ~]$ scp -r /usr/local/hbase root@slave1:/usr/local
root@slave1's password:
LICENSE.txt          100%  136KB  32.8MB/s   00:00
NOTICE.txt           100%  561KB  46.8MB/s   00:00
LEGAL                100%  262   220.2KB/s   00:00
package-frame.html   100%  661   867.4KB/s   00:00
```

图6-21　在master节点复制hbase文件夹至slave1

（2）配置目录权限

在slave1、slave2节点修改hbase文件夹的所属者为hadoop用户，如图6-22所示。

chown -R hadoop:hadoop /usr/local/hbase

```
[root@slave1 local]# chown -R hadoop:hadoop hbase
[root@slave1 local]# ll
总用量 4
drwxr-xr-x.  2 root   root      6 8月   9 2021 bin
drwxr-xr-x.  2 root   root      6 8月   9 2021 etc
drwxr-xr-x.  2 root   root      6 8月   9 2021 games
drwxr-xr-x. 12 hadoop hadoop 4096 10月 22 03:47 hadoop
drwxr-xr-x.  7 hadoop hadoop  182 10月 22 04:03 hbase
```

图6-22　在slave1节点设置hbase文件夹权限

（3）配置PATH

使用vim编辑器编辑/etc/profile：

vim /etc/profile

在末行增加如下内容：

export HBASE_HOME=/usr/local/hbase
export PATH=$PATH:$HBASE_HOME/bin

5．启动与关闭HBase

（1）启动HBase

启动HBase前，先启动Hadoop的全部进程。

> ⚠ 注意
> 启动关闭HBase的顺序是：启动Hadoop → 启动HBase → 关闭HBase → 关闭Hadoop。

> **说明**
>
> Hadoop集群启动成功时，在master节点看到NameNode、DataNode、Secondary NameNode、NodeManager、ResourceManager进程；在slave节点看到DataNode、NodeManager进程。

```
$ start-all.sh
```

启动Hadoop后，各节点进程如图6-23所示。

图6-23 启动HBase

a）master节点 b）slave节点

Hadoop成功启动后，启动HBase。

```
$ start-hbase.sh
```

启动HBase后，各节点进程如图6-24所示。

> **说明**
>
> HBase集群启动成功时，在master节点看到HMaster、HQuorumPeer、HRegionServer三个进程；在slave节点看到HQuorumPeer、HRegionServer两个进程。进程中，HMaster对应Master服务器；HQuorumPeer对应HBase自带的ZooKeeper；HRegionServer对应Region服务器。

图6-24 启动完全分布式HBase

a）master节点 b）slave节点

（2）查看HBase的Web界面

打开浏览器，在地址栏输入"localhost:16010"，查看HBase的Web界面，如图6-25所示。

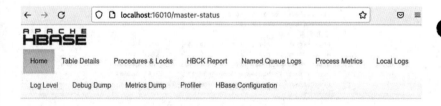

> **说明**
>
> 利用HBase的Web界面可以查看HBase集群运行状态，读者可自行查看并学习。

图6-25 HBase的Web界面

（3）关闭HBase

```
$ stop-hbase.sh
```

项目6 部署与使用 HBase

扫码看视频

任务3　利用Shell操作HBase

任务描述

客户要求在HBase分布式数据库中创建学生信息表，见表6-5。王工利用Shell命令实现客户要求，并为客户演示数据库表的各类操作。

表6-5　学生信息表

学号	姓名	性别	年龄	专业	课程成绩		
					计算机（必修）	数学（选修）	英语（选修）
95001	Zhangsan	M	22	CS	95	—	75
95002	Lisi	F	20	CS	80	80	—

任务分析

1. 任务目标

1）学会启动和停止HBase。

2）学会启动和退出HBase Shell环境。

3）学会在HBase Shell环境下，完成表的创建、删除、查看等。

4）学会在HBase Shell环境下，对表数据添加、查看、读取等。

2. 任务环境

操作系统：CentOS Stream 9（预装分布式Hadoop+HBase）

软件版本：Java 1.8.0、Hadoop 3.3.4、HBase 2.4.14

3. 任务导图

任务导图如图6-26所示。

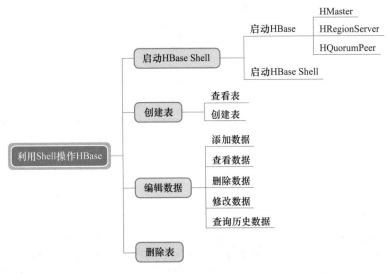

图6-26　任务导图

> ⚠ 注意
>
> 启动关闭 HBase 的顺序是：启动 Hadoop → 启动 HBase → 关闭 HBase → 关闭 Hadoop。

任务实施

1. 启动HBase Shell

（1）启动HBase

启动Hadoop与HBase，如图6-27所示。

```
$ start-all.sh
$ start-hbase.sh
```

启动完成后，可以看到HBase的所有进程。

```
[hadoop@master ~]$ jps
11287 ResourceManager
11576 Jps
11065 SecondaryNameNode
10859 DataNode
10732 NameNode
11390 NodeManager
[hadoop@master ~]$ start-hbase.sh
slave2: running zookeeper, logging to /usr/local/hbase/bin/../logs/hbase-hadoop-zookeeper-slave2.out
slave1: running zookeeper, logging to /usr/local/hbase/bin/../logs/hbase-hadoop-zookeeper-slave1.out
master: running zookeeper, logging to /usr/local/hbase/bin/../logs/hbase-hadoop-zookeeper-master.out
running master, logging to /usr/local/hbase/logs/hbase-hadoop-master-master.out
slave2: running regionserver, logging to /usr/local/hbase/bin/../logs/hbase-hadoop-regionserver-slave2.out
slave1: running regionserver, logging to /usr/local/hbase/bin/../logs/hbase-hadoop-regionserver-slave1.out
master: running regionserver, logging to /usr/local/hbase/bin/../logs/hbase-hadoop-regionserver-master.out
[hadoop@master ~]$ jps
12241 HRegionServer
12084 HMaster
12341 Jps
11287 ResourceManager
11991 HQuorumPeer
11065 SecondaryNameNode
10859 DataNode
10732 NameNode
11390 NodeManager
```

图6-27　启动HBase

（2）启动HBase Shell

启动HBase Shell，如图6-28所示。

```
$ hbase shell
```

```
[hadoop@master ~]$ hbase shell
2022-10-22 08:49:45,911 WARN  [main] util.NativeCodeLoader: Unable to load native-hadoop library for your platform... using builtin-java classes where applicable
HBase Shell
Use "help" to get list of supported commands.
Use "exit" to quit this interactive shell.
For Reference, please visit: http://hbase.apache.org/2.0/book.html#shell
Version 2.4.14, r2e7d75a892000071a7479b2f668c4db7a241be3f, Tue Aug 23 23:33:09 UTC 2022
Took 0.1146 seconds
hbase:001:0>
```

图6-28　HBase Shell启动成功界面

> 📝 笔记
>
> 若退出 HBase Shell 环境，只要输入 exit; 即可返回工作目录

2. 创建表

（1）查看表

查看当前数据库中的所有数据表，如图6-29所示。

```
> list
```

> ⚠ 注意
>
> 在关系型数据库（如 MySQL）中，需要首先创建数据库，然后创建表，但是在 HBase 数据库中，不需要创建数据库，只要直接创建表就可以。

```
hbase:001:0> list
TABLE
0 row(s)
Took 0.9666 seconds
=> []
hbase:002:0>
```

图6-29　查看当前库中已有的表

（2）创建表

创建一张student表，包括字段为：姓名（name）、性别（sex）、年龄（age）、专业（major）、课程成绩（score），如图6-30所示。

> create 'student', 'name', 'sex', 'age', 'major', 'score'

```
hbase:002:0> create 'student','name','sex','age','major',
'score'
Created table student
Took 0.8033 seconds
=> Hbase::Table - student
```

图6-30　创建student数据表

创建完student表后，可通过d<esc>ribe命令查看student表的基本信息，如图6-31所示。

> d<esc>ribe 'student'

```
hbase:006:0> describe 'student'
Table student is ENABLED
student
COLUMN FAMILIES DESCRIPTION
{NAME => 'age', BLOOMFILTER => 'ROW', IN_MEMORY => 'false', VERSIONS => '1', KEE
P_DELETED_CELLS => 'FALSE', DATA_BLOCK_ENCODING => 'NONE', COMPRESSION => 'NONE'
, TTL => 'FOREVER', MIN_VERSIONS => '0', BLOCKCACHE => 'true', BLOCKSIZE => '655
36', REPLICATION_SCOPE => '0'}

{NAME => 'major', BLOOMFILTER => 'ROW', IN_MEMORY => 'false', VERSIONS => '1', K
EEP_DELETED_CELLS => 'FALSE', DATA_BLOCK_ENCODING => 'NONE', COMPRESSION => 'NON
E', TTL => 'FOREVER', MIN_VERSIONS => '0', BLOCKCACHE => 'true', BLOCKSIZE => '6
5536', REPLICATION_SCOPE => '0'}

{NAME => 'name', BLOOMFILTER => 'ROW', IN_MEMORY => 'false', VERSIONS => '1', KE
EP_DELETED_CELLS => 'FALSE', DATA_BLOCK_ENCODING => 'NONE', COMPRESSION => 'NONE
', TTL => 'FOREVER', MIN_VERSIONS => '0', BLOCKCACHE => 'true', BLOCKSIZE => '65
536', REPLICATION_SCOPE => '0'}

{NAME => 'score', BLOOMFILTER => 'ROW', IN_MEMORY => 'false', VERSIONS => '1', K
EEP_DELETED_CELLS => 'FALSE', DATA_BLOCK_ENCODING => 'NONE', COMPRESSION => 'NON
E', TTL => 'FOREVER', MIN_VERSIONS => '0', BLOCKCACHE => 'true', BLOCKSIZE => '6
5536', REPLICATION_SCOPE => '0'}

{NAME => 'sex', BLOOMFILTER => 'ROW', IN_MEMORY => 'false', VERSIONS => '1', KEE
P_DELETED_CELLS => 'FALSE', DATA_BLOCK_ENCODING => 'NONE', COMPRESSION => 'NONE'
, TTL => 'FOREVER', MIN_VERSIONS => '0', BLOCKCACHE => 'true', BLOCKSIZE => '655
36', REPLICATION_SCOPE => '0'}
```

图6-31　查看表信息

使用list命令查看当前HBase数据库中的表，如图6-32所示，可以看到新建的student表。

> list

```
hbase:007:0> list
TABLE
student
1 row(s)
Took 0.0484 seconds
=> ["student"]
```

图6-32　查看数据库中表列表

> **⚠ 注意**
>
> 对于HBase而言，在创建HBase表时，不需要自行创建行键，系统会默认一个属性作为行键，通常是把 put 命令操作中跟在表名后的第一个数据作为行键。

> **ⓘ 扩展**
>
> 简单了解表信息中每个列族的信息描述：
>
> NAME：列族名称。
> BLOOMFILTER：布隆过滤器。
> IN_MEMORY：是否在内存中。
> VERSIONS：保留版本个数。
> KEEP_DELETED_CELLS：保留被删除的行。
> DATA_BLOCK_ENCODING：数据块编码。
> COMPRESSION：压缩。
> TTL：数据存活时间。
> MIN_VERSIONS：最小版本。
> BLOCKCACHE：是否开启块缓存。
> BLOCKSIZE：缓存块大小。
> REPLICATION_SCOPE：复制范围。

3. 编辑数据

（1）添加数据

依次插入Zhangsan的数据，如图6-33所示。

> put 'student','95001','name','Zhangsan'
> put 'student','95001','sex','male'
> put 'student','95001','age','22'
> put 'student','95001','major','CS'
> put 'student','95001','score:computer','95'
> put 'student','95001','score:English','75'

```
hbase:010:0> put 'student','95001','name','Zhangsan'
Took 0.3677 seconds
hbase:011:0> put 'student','95001','sex','male'
Took 0.0094 seconds
hbase:012:0> put 'student','95001','age','22'
Took 0.0104 seconds
hbase:013:0> put 'student','95001','major','CS'
```

图6-33 添加数据

笔记

HBase 使用 put 命令添加数据，一次只能为一个表的一行数据的一列（也就是一个单元格）添加一个数据。所以，直接用 Shell 命令插入数据效率很低，在实际应用中，一般都是利用编程操作数据。

说明

put 命令会为 student 表添加学号为 '95001'、姓名为 'Zhangsan' 的一个单元格数据，其行键为 95001。即系统默认把跟在表名 student 后面的第一个数据作为行键。

（2）查看数据

使用get命令查看某一行数据，如图6-34所示。

> get 'student','95001'

```
hbase:020:0> get 'student','95001'
COLUMN                CELL
 age:                 timestamp=2022-08-15T08:10:19.804, value=22
 major:               timestamp=2022-08-15T08:10:31.432, value=CS
 name:                timestamp=2022-08-15T08:09:27.331, value=Zhangsan
 score:English       timestamp=2022-08-15T08:12:54.601, value=75
 score:computer      timestamp=2022-08-15T08:11:34.001, value=95
 sex:                 timestamp=2022-08-15T08:10:02.898, value=male
1 row(s)
Took 0.1476 seconds
```

图6-34 get命令查看数据

笔记

HBase 中有两个用于查看数据的命令：
get 命令：用于查看表的某一行数据；
scan 命令：用于查看某个表的全部数据。

按同样的方式，添加表中第二行内容后，使用scan查看表内全部数据，如图6-35所示。

> scan 'student'

```
hbase:030:0> scan 'student'
ROW                   COLUMN+CELL
 95001                column=age:, timestamp=2022-08-15T08:10:19.804, value=22
 95001                column=major:, timestamp=2022-08-15T08:10:31.432, value=CS
 95001                column=name:, timestamp=2022-08-15T08:09:27.331, value=Zhangsan
 95001                column=score:English, timestamp=2022-08-15T08:12:54.601, value=75
 95001                column=score:computer, timestamp=2022-08-15T08:11:34.001, value=95
 95001                column=sex:, timestamp=2022-08-15T08:10:02.898, value=male
 95002                column=age:, timestamp=2022-08-15T08:21:25.044, value=20
 95002                column=major:, timestamp=2022-08-15T08:21:35.786, value=CS
 95002                column=name:, timestamp=2022-08-15T08:21:07.042, value=Lisi
 95002                column=score:computer, timestamp=2022-08-15T08:21:50.648, value=80
 95002                column=score:math, timestamp=2022-08-15T08:22:00.251, value=80
 95002                column=sex:, timestamp=2022-08-15T08:21:15.064, value=F
2 row(s)
Took 0.0203 seconds
```

图6-35 scan命令查看数据

至此数据已添加完成，后面的内容将进行数据的删除、修改等操作。

（3）删除数据

删除Zhangsan的英语成绩，如图6-36所示。

> delete 'student','95001','score:English'

```
hbase:031:0> delete 'student','95001','score:English'
Took 0.0184 seconds
hbase:032:0> get 'student','95001'
COLUMN                CELL
 age:                 timestamp=2022-08-15T08:10:19.804, value=22
 major:               timestamp=2022-08-15T08:10:31.432, value=CS
 name:                timestamp=2022-08-15T08:09:27.331, value=Zhangsan
 score:computer       timestamp=2022-08-15T08:11:34.001, value=95
 sex:                 timestamp=2022-08-15T08:10:02.898, value=male
1 row(s)
Took 0.0198 seconds
```

图6-36 删除数据

使用deleteall命令删除student表中的95001行的全部数据，如图6-37所示。

> deleteall 'student','95001'

```
hbase:033:0> deleteall 'student','95001'
Took 0.0051 seconds
hbase:034:0> get 'student','95001'
COLUMN                CELL
0 row(s)
Took 0.0101 seconds
```

图6-37 删除一行数据

（4）修改数据

修改Lisi的年龄为21岁，如图6-38所示。

>put 'student','95002','age','21'

```
hbase:035:0> put 'student','95002','age','21'
Took 0.0101 seconds
hbase:036:0> get 'student','95002'
COLUMN                CELL
 age:                 timestamp=2022-08-15T08:28:31.147, value=21
 major:               timestamp=2022-08-15T08:21:35.786, value=CS
 name:                timestamp=2022-08-15T08:21:07.042, value=Lisi
 score:computer       timestamp=2022-08-15T08:21:50.648, value=80
 score:math           timestamp=2022-08-15T08:22:00.251, value=80
 sex:                 timestamp=2022-08-15T08:21:15.064, value=F
1 row(s)
Took 0.0181 seconds
```

图6-38 修改数据

📝 **说明**

查询时，默认情况下会显示当前最新版本的数据，如果要查询历史数据，需要指定查询的历史版本数。但是根据student表中age列族的描述，当前仅保留一个版本，故修改数据后，看不到旧版本数据。

（5）查询历史数据

修改age列族的VERSIONS属性值为3，如图6-39所示。

>disable 'student'
>alter 'student',NAME=>'age',VERSIONS=>3
>enable 'student'
>d<esc>ribe 'student'

📝 **笔记**

修改表属性时，需要先设置表不可用（disable表），再修改属性，修改完成后，重新使用表（enable表）。

```
hbase:053:0> alter 'student',NAME=>'age',VERSIONS=>3
Updating all regions with the new schema...
All regions updated.
Done.
Took 1.1163 seconds

hbase:054:0> enable 'student'
Took 0.6212 seconds

hbase:055:0> describe 'student'
Table student is ENABLED

student

COLUMN FAMILIES DESCRIPTION

{NAME => 'age', BLOOMFILTER => 'ROW', IN_MEMORY => 'false', VERSIONS => '3'
, KEEP_DELETED_CELLS => 'FALSE', DATA_BLOCK_ENCODING =>
 'NONE', COMPRESSION => 'NONE', TTL => 'FOREVER', MIN_VERSIONS => '0', BLOC
KCACHE => 'true', BLOCKSIZE => '65536', REPLICATION_SCO
PE => '0'}
```

图6-39 修改列族属性

说明

修改age列族属性后，age列族可以存储3个历史值。因此，查看age列族数据时，可以使用VERSIONS参数查看最近3个历史版本值。

依次修改Lisi年龄为21、22、23，并查看历史数据，如图6-40所示。

```
> put 'student','95002','age','21'
> put 'student','95002','age','22'
> put 'student','95002','age','23'
> get 'student','95002',{COLUMN=>'age',VERSIONS=>5}
```

```
hbase:067:0> put 'student','95002','age','21'
Took 0.0024 seconds
hbase:068:0> put 'student','95002','age','22'
Took 0.0036 seconds
hbase:069:0> put 'student','95002','age','23'
Took 0.0027 seconds
hbase:070:0> get 'student','95002',{COLUMN=>'age',VERSIONS=>3}
COLUMN                CELL
 age:                 timestamp=2022-08-15T08:48:45.996, value=23
 age:                 timestamp=2022-08-15T08:48:44.082, value=22
 age:                 timestamp=2022-08-15T08:48:41.825, value=21
1 row(s)
Took 0.0167 seconds
```

图6-40　查看历史数据

4. 删除表

删除表前，需要设置student表不可用，如图6-41所示。

```
> disable 'student'
> drop 'student'
```

```
hbase:071:0> list
TABLE
student
1 row(s)
Took 0.0070 seconds
=> ["student"]
hbase:072:0> disable 'student'
Took 0.3223 seconds
hbase:073:0> drop 'student'
Took 0.3720 seconds
hbase:074:0> list
TABLE
0 row(s)
Took 0.0026 seconds
=> []
```

图6-41　删除表

拓展学习

1. Region服务器的工作原理

（1）Region

Region是HBase中负载均衡和数据分发的基本单位，一个Region默认大小为100～200MB。

在HBase中，如图6-42所示，一张HBase表中，行键按照字典序排序，然后按照行键划分成多个Region。

扫码看视频

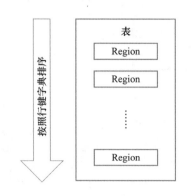

图6-42　HBase表划分成多个Region

如图6-43所示，一张表一开始只有一个Region，随着表的内容增加，超出了一个Region的限定大小，则Region会进一步分裂，最终分裂成多个新的Region。

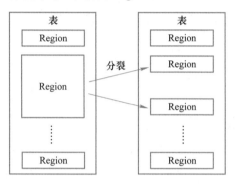

图6-43　Region表裂变

存储了表数据的Region会被Master服务器分配到某一个Region服务器上，一个Region服务器可以管理10～1000个Region。

（2）Region的定位寻址

用户通过访问Region，得到Region内的用户数据。客户端在访问请求数据时，并不是直接从Master主服务器上读取数据，而是通过ZooKeeper来获得Region的位置信息。如图6-44所示，采用三级寻址的方式：访问根地址（-ROOT-表），再访问存放Region块的元数表（.META表），最后获取数据。

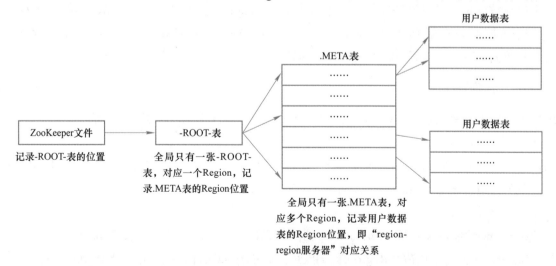

图6-44　客户端访问HBase三级寻址

（3）Region服务器结构

Region服务器结构如图6-45所示。

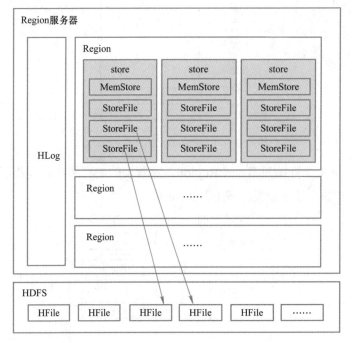

图6-45　Region服务器结构

每一个Region服务器内需要管理多个Region和一个HLog对象。HLog对象用于记录当前Region服务器内所有Region的更新操作。每一个Region由多个Store组成，每个Store对应一个Region中的列族。Store中的MemStore是内存缓存，实现快速数据更新；StoreFile是文件存储，以HFile的形式存储于底层HDFS文件系统中，实现数据持久化。

（4）Region读写过程

当用户读取数据时，通过三级寻址的方式得到用户数据所在的Region服务器位置，Region服务器定位到相应的Region，根据列族定位到Store。首先从Memstore中读取，如果读取不到，则到StoreFile中读取。

当用户写数据时，根据行健的值，会被分配到相应Region服务器去执行。在Region服务器内，按照列族，找到相应的Store。写操作会分别记录在列族对应的Store的MemStore及HLog中。为了保证服务器意外断电时数据不丢失，必须保证HLog中的数据写入操作的准确性。

2．HLog工作原理

HBase采用HLog保证系统出错时能恢复到正确状态。下面介绍HLog的工作原理。

（1）数据写入过程

HBase系统为每个Region服务器配置了一个HLog文件。它是一种预写式日志（Write Ahead Log），用户更新数据必须首先写入HLog日志，然后才能写入MemStore缓存。

（2）数据恢复过程

1）ZooKeeper会实时监测每个Region服务器的状态，当发生故障时，会通知Master。

2）Master首先会处理该故障Region服务器上遗留的HLog文件，这个遗留的HLog文件中包含了来自多个Region对象的日志记录。

3）系统会根据每条日志，记录所属的Region对象，对HLog数据进行拆分，分别放到相应Region对象的目录下。然后将失效的Region重新分配到可用的Region服务器中，并把与该Region对象相关的HLog日志记录也发送给相应的Region服务器。

4）Region服务器领取到分配给自己的Region对象以及与之相关的HLog日志记录后，会重新执行一遍日志记录中的各种操作，把日志记录中的数据写入到MemStore缓存中。然后刷新到磁盘的StoreFile文件中，完成数据恢复。

HLog的优点：提高对表的写操作性能；缺点：恢复时需要分拆日志。

3. Store工作原理

Store是Region服务器的核心，工作原理如图6-46所示。MemStore内的数据会定时刷新生成StoreFile，多个StoreFile可以合并成一个StoreFile，提高数据查询效率。当一个Region太大时，需要拆分Region，同时需要拆分Storefile，又触发分裂操作，一个父Region被分裂成两个子Region，替代原父Region继续工作。

图6-46　Store工作原理

项目小结

本项目围绕"部署与使用HBase"典型工作任务，以HBase的交付部署作为工作背景，讲解了HBase相关基本概念，并展开三个任务，学习如何部署分布式HBase，并基于分布式HBase，使用Shell实现HBase数据表的维护，伪分布式与全分布式部署过程的对比如图6-47所示。在完成任务后，开展扩展知识学习，包括HBase底层运行原理，进一步丰富运维工程师的知识储备。

	任务1：部署伪分布模式	任务2：部署全分布模式	
环境配置	安装配置伪分布模式Hadoop	安装配置完全分布模式Hadoop 关闭防火墙（三台主机）	
安装HBase	下载、解压 配置目录权限 配置PATH	下载、解压 配置目录权限 配置PATH	master节点
配置参数	配置hbase-env.sh 配置hbase-site.xml	配置hbase-env.sh 配置hbase-site.xml 配置regionservers 配置版本兼容性	master节点
同步配置		同步Hadoop至slave节点 配置目录权限	slave节点
启动与关闭	启动Hadoop 启动HBase 关闭HBase 关闭Hadoop	启动Hadoop（三台主机同时启动） 启动HBase（三台主机同时启动） 关闭HBase（三台主机同时关闭） 关闭Hadoop（三台主机同时关闭）	

图6-47　伪分布与完全分布模式配置流程横向对比总结

实战强化

1. 由于一些未知原因，HBase集群在关闭过程中没有关闭HRegionServer进程，如图6-48所示，如需关闭该进程，应使用什么指令？

```
[hadoop@master hadoop]$ stop-hbase.sh
no hbase master found
master: running zookeeper, logging to /usr/local/hbase/bin/../logs/
hbase-hadoop-zookeeper-master.out
slave2: running zookeeper, logging to /usr/local/hbase/bin/../logs/
hbase-hadoop-zookeeper-slave2.out
slave1: running zookeeper, logging to /usr/local/hbase/bin/../logs/
hbase-hadoop-zookeeper-slave1.out
master: stopping zookeeper.
slave2: stopping zookeeper.
slave1: stopping zookeeper.
[hadoop@master hadoop]$ jps
10800 DataNode
11042 SecondaryNameNode
10631 NameNode
11416 NodeManager
13097 Jps
11277 ResourceManager
12350 HRegionServer
```

图6-48　HRegionServer未正常关闭

2. 运维人员想从HBase中删除teacher表，运行结果如图6-49所示。

```
hbase:002:0> drop 'teacher'

ERROR: Table teacher is enabled. Disable it first.

For usage try 'help "drop"'

Took 2.2449 seconds
hbase:003:0>
```

图6-49　删除teacher表报错图

试分析系统报错的原因，并给出解决方案。

项目 7　部署与使用 Hive

●项目概述

随着数据规模的增大和数据形式的复杂化，A公司之前部署的基于Hadoop的大数据平台项目中MapReduce的编写过于复杂，为提高客户大数据平台使用效率，A公司为客户在Hadoop集群基础上，新增部署Hive。现公司委派大数据运维工程师王工前往客户的中心机房对平台进行Hive集群部署，并为客户演示利用Hive实现数据导入。

本项目针对"部署与使用Hive"的典型工作任务，由易到难，依次完成四个工作任务：部署本地模式Hive、部署远程模式Hive、利用Hive实现数据导入、利用Hive实现词频统计。

在开展任务前，需要掌握必要的理论知识：什么是Hive？Hive与Hadoop的区别是什么？三种部署模式分别指的是什么？Hive与传统数据库的区别是什么？Hive的表类型及数据类型是如何定义的？

完成任务后，进一步了解与Hive相关的知识，加深对Hive的理解。

●学习目标

1. 了解Hive简介、生态系统、安装方式。
2. 理解部署Hive与MySQL的关系。
3. 掌握Hive基本语法。
4. 掌握Hive数据类型。
5. 能按照步骤完成Hive的部署（本地模式、远程模式）。
6. 熟练应用Hive实现数据库操作。
7. 了解Hive的体系架构。
8. 了解Hive的工作原理。

●思维导图

项目思维导图如图7-1所示。

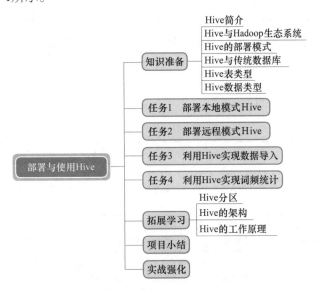

图7-1　项目思维导图

知识准备

1. Hive简介

Hive是基于Hadoop的一个数据仓库工具，可以将结构化的数据文件映射为一张数据库表，并提供简单的SQL查询功能，可以将SQL语句转换为MapReduce任务运行。其优点是学习成本低，可以通过类SQL语句快速实现简单的MapReduce统计，不必开发专门的MapReduce应用，非常适合数据仓库的统计分析。

2. Hive与Hadoop生态系统

Hadoop生态系统如图7-2所示。

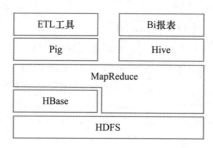

图7-2　Hadoop生态系统

（1）Hive依赖于HDFS存储数据

HDFS作为高可靠性的底层存储，用来存储海量数据。

（2）Hive依赖于MapReduce处理数据

MapReduce对这些海量数据进行处理，实现高性能计算，用HiveQL语句编写的处理逻辑最终均要转化为MapReduce任务来运行。

（3）Pig可以作为Hive的替代工具

Pig是一种数据流语言和运行环境，适合用于Hadoop和MapReduce平台上查询半结构化数据集。其常用于ETL过程的一部分，即将外部数据装载到Hadoop集群中，然后转换为用户期待的数据格式。

（4）HBase提供数据的实时访问

HBase是一个面向列的、分布式的、可伸缩的数据库，它可以提供数据的实时访问功能，而Hive只能处理静态数据，主要是BI报表数据，所以HBase与Hive的功能是互补的，它实现了Hive不能提供的功能。

3. Hive的部署模式

Hive有三种部署模式，见表7-1。

表7-1　Hive的部署模式

部署模式	具体说明
内嵌模式	使用内嵌的Derby数据库来存储元数据，不需要额外启动Metastore服务。 数据库和Metastore服务都嵌入在主Hive Server进程中。配置简单，但一次只能连接一个客户端，适用于测试环境，不适用于生产环境
本地模式	采用独立数据库（MySQL）存储元数据，Hive客户端和Metastore服务在同一台服务器中启动，Hive客户端通过连接本地的Metastore服务获取元数据信息。支持元数据共享，并且支持本地多会话连接
远程模式	与本地模式一样，采用独立数据库（MySQL）存储元数据，不同的是Hive客户端和Metastore服务在不同的服务器启动，Hive客户端通过远程连接Metastore服务获取元数据信息。支持元数据共享，并且支持远程多会话连接。在生产环境中，建议用远程模式来配置Hive Metastore

4. Hive与传统数据库

Hive与传统数据库的区别见表7-2。

表7-2　Hive与传统数据库的区别

	传统数据库	Hive
数据插入	同时支持导入单条数据和批量数据	仅支持批量导入数据。Hive主要用来支持大规模数据集上的数据仓库应用程序的运行，常见操作是全表扫描，单条插入功能对Hive并不实用
数据更新	支持	不支持。数据仓库中存放的是静态数据
索引	支持	Hive 0.7版本后支持索引，但只提供有限的索引功能，使用户可以在某些列上创建索引来加速一些查询操作，Hive中给一个表创建的索引数据被保存在另外的表中
分区	提供分区功能来改善大型表以及具有各种访问模式的表的可伸缩性、可管理性和提高数据库效率	支持，Hive表组织成分区的形式，根据分区列的值对表进行粗略的划分，使用分区可以加快数据的查询速度
执行延迟	少于一秒	Hive的延迟比较高，HiveQL语句的延迟会达到分钟级
扩展性	很难横向扩展，纵向扩展的空间也很有限	Hive的开发环境是基于集群的，具有较好的可扩展性

5. Hive表类型

Hive 管理数据的方式主要包括如下几种：Managed Table（内部表）、External Table（外部表）、Partition（分区）和Bucket（桶），具体见表7-3。

表7-3　Hive表类型

表 类 型	具 体 描 述
内部表	与关系型数据库中的表在概念上很类似，每个表在 HDFS 中都有相应的目录存储表的数据，由 Hive 管理表和与表相关的数据。当表定义被删除的时候，该表对应的所有数据包括元数据和表数据都会被删除
外部表	与内部表相似，但是其数据不是存储在自己表所属的目录中，而是存储到别处，这样的好处是，如果要删除这个外部表，该外部表所指向的数据是不会被删除的，它只会删除外部表对应的元数据（metadata）
分区	表的每一个分区对应表目录下相应的一个子目录，所有分区的数据都是存储在对应的子目录中
桶	对指定的列值计算其 hash，根据hash值切分数据，目的是为了并行，每一个桶对应一个文件（注意和分区的区别）

6. Hive数据类型

（1）基本数据类型

TINYINT（1B）、SMALLINT（2B）、INT（4B）、BIGINT（8B）、FLOAT（4B）、DOUBLE（8B）、BOOLEAN（true/false）、STRING（字符序列）。

Hive也是由Java编写的，所以Hive的基本数据类型都是对Java中的接口的实现，这些基本的数据类型和Java的基本数据类型是一一对应的。

（2）复杂数据类型

STRUCT：结构体，通过属性名访问属性值。格式为STRUCT（first string，last string），第一个元素可以通过"列名.first"访问。

任务1　部署本地模式Hive

任务描述

王工在前往客户现场前，为了确保顺利实现Hive的集群部署，先在自己公司的单机环境下进行本地模式部署，并测试Hive的基础功能。

任务分析

1. 任务目标

1）理解Hive的工作原理及体系架构。

2）掌握Hive本地模式部署。

3）能够简单运行Hive程序。

2. 任务环境

操作系统：CentOS Stream 9（预装分布式Hadoop）

软件版本：Java 1.8.0、MySQL 5.7.24、Hadoop3.3.4、Hive 3.1.2

3. 任务导图

任务导图如图7-3所示。

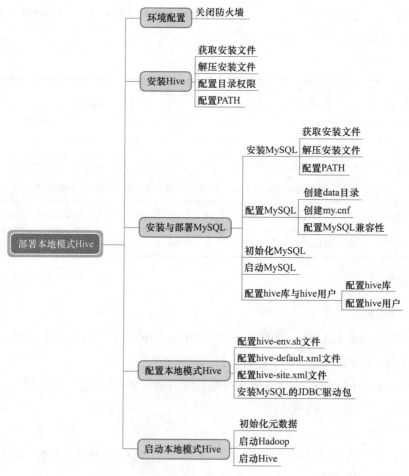

图7-3　任务导图

任务实施

1. 环境配置

运行指令，关闭所有节点的防火墙，并查看防火墙状态，确认防火墙已关闭。

```
# systemctl stop firewalld
# systemctl status firewalld
```

> **说明**
> 如果在伪分布模式Hadoop下部署，可不关闭防火墙。如果在全分布模式Hadoop下，为保证Hadoop正常运行，需要关闭防火墙。

2. 安装Hive

（1）获取安装文件

访问Hive官网（https://hive.apache.org/）下载安装文件apache-hive-3.1.2-bin.tar.gz，如图7-4所示。

```
# cd /usr/local
# wget https://downloads.apache.org/hive/hive-3.1.2/apache-hive-3.1.2-bin.tar.gz
```

```
[root@localhost lib64]# cd /usr/local
[root@localhost local]# wget https://downloads.apache.org/hive/hive-3.1.2/apache-hive-3.1.2-bin.tar.gz
--2022-08-21 07:18:51--  https://downloads.apache.org/hive/hive-3.1.2/apache-hive-3.1.2-bin.tar.gz
正在解析主机 downloads.apache.org (downloads.apache.org)... 198.18.9.232
正在连接 downloads.apache.org (downloads.apache.org)|198.18.9.232|:443... 已连接。
已发出 HTTP 请求，正在等待回应... 200 OK
长度：278813748 (266M) [application/x-gzip]
正在保存至: "apache-hive-3.1.2-bin.tar.gz"
```

图7-4 下载Hive安装文件

> **说明**
> 本书在本任务中的截图来自于Hadoop伪分布环境。

> **说明**
> 随着版本更新，当前版本号的下载地址可能发生变化，请读者自行到官网获取相应下载地址。

（2）解压安装文件

```
# tar -zxf apache-hive-3.1.2-bin.tar.gz
```

更改目录名称为hive：

```
# mv apache-hive-3.1.2-bin hive
```

（3）配置目录权限

将目录的所有者改为hadoop用户，如图7-5所示。

```
# chown -R hadoop:hadoop hive
```

```
[root@localhost local]# mv apache-hive-3.1.2-bin hive
[root@localhost local]# chown -R hadoop:hadoop hive
[root@localhost local]# ls -ld hive
drwxr-xr-x. 10 hadoop hadoop 184 8月  21 07:21 hive
```

图7-5 重命名Hive文件夹并更改目录权限

> **说明**
> 如果通过wget指令获取安装包太慢，可以使用本书提供的安装包。使用scp指令，从Windows传输文件至Linux中。

> **说明**
> 本任务需要在完成Hadoop分布式部署的环境下进行，伪分布或全分布模式均可。

（4）配置PATH

使用vim编辑器编辑/etc/profile。

```
# vim /etc/profile
```

在打开的profile文件中，添加Hive相关路径。

在末行增加如下内容，如图7-6所示。

```
export HIVE_HOME=/usr/local/hive
export PATH=$PATH:$HIVE_HOME/bin
```

```
export JAVA_HOME=/usr/lib/jvm/java-1.8.0-openjdk-1.8.0.345.b01-2.el9.x86_64/jre
export PATH=$PATH:$JAVA_HOME

export HADOOP_HOME=/usr/local/hadoop
export PATH=$PATH:$HADOOP_HOME/bin:$HADOOP_HOME/sbin

export HIVE_HOME=/usr/local/hive
export PATH=$PATH:$HIVE_HOME/bin
```

图7-6 配置环境变量

扫码看视频

📝 **笔记**

为什么安装 MySQL？

Hive 的数据由两部分组成：数据文件和元数据。元数据用于存放 Hive 库的基础信息，它存储在关系数据库中，如 MySQL、Derby。这里采用 MySQL 数据库保存 Hive 的元数据，而不是采用 Hive 自带的 Derby 来存储元数据，因此，需要安装 MySQL 数据库。

⚠ **注意**

如果在 CentOS 7 中安装，需要先卸载系统自带的 MariaDB。使用指令查询已安装的 mariadb 软件包：rpm -qa | grep mariadb。卸载 mariadb 软件包：rpm -e -nodeps 包名。还需要另外安装 libaio 包，使用指令：yum install libaio。

配置完成后，按<ESC>键，输入:wq，保存并退出。

使用 source 命令使新配置生效。

```
# source /etc/profile
```

3. 安装与部署 MySQL

（1）安装 MySQL

1）获取安装文件。

从 MySQL 官网（https://downloads.mysql.com/archives/community/）下载安装文件 mysql-*.*.*.tar.gz，例如 mysql-5.7.24-linux-glibc2.12-x86_64.tar.gz，如图 7-7 所示。

```
# cd /usr/local
#wget https://downloads.mysql.com/archives/get/p/23/file/mysql-5.7.24-linux-glibc2.12-x86_64.tar.gz
```

图 7-7　安装 MySQL

2）解压安装文件。

解压 MySQL 安装包，并重命名为 mysql。

```
# tar -zxf mysql-5.7.24-linux-glibc2.12-x86_64.tar.gz
# mv mysql-5.7.24-linux-glibc2.12-x86_64 mysql
```

3）配置 PATH。

配置 MySQL 安装路径至 PATH 变量。

```
# vi /etc/profile
```

添加如下内容，如图 7-8 所示。

export MYSQL_HOME=/usr/local/mysql
export PATH=$PATH:$MYSQL_HOME/bin

图 7-8　配置 PATH

⚠ **注意**

不要忘记使环境变量生效。source 命令的功能就是使其生效。若使用 source 命令后，发现之前配置的环境变量存在拼写等错误，则在终端输入 export PATH=/usr/bin:/usr/sbin:/bin/sbin，按 <Enter> 键后重新进入配置文件。

按<ESC>键，然后输入:wq，保存退出。

使用source命令使环境变量生效。

\# source /etc/profile

（2）配置MySQL

1）创建data目录。

进入mysql路径创建data目录，如图7-9所示。

⚠ 注意

未创建data目录将可能导致初始化失败。

\# cd /usr/local/mysql
\# mkdir data

```
[root@localhost ~]# cd /usr/local/mysql
[root@localhost mysql]# mkdir data
```

图7-9　创建data目录

2）创建my.cnf。

在系统/etc路径下，创建并配置my.cnf文件。

📝 笔记

Linux下的大部分系统配置都在/etc/目录下。

📝 笔记

my.cnf内容解析见表7-4。

\# cd /etc
\# vi my.cnf

在my.cnf中添加如下内容，如图7-10所示。

```
[mysqld]
basedir=/usr/local/mysql
datadir=/usr/local/mysql/data
socket=/tmp/mysql.sock
lower_case_table_names=1
user=root
#default-character-set=utf8
character-set-server=utf8
# Disabling symbolic-links is recommended to prevent assorted security risks
symbolic-links=0
[client]
default-character-set=utf8
[mysqld_safe]
log-error=/var/log/mysqld.log
pid-file=/var/run/mysqld/mysqld.pid
```

表7-4　my.cnf内容解析

内容	解析
basedir	MySQL的安装目录
datadir	MySQL的数据存储目录
socket	为MySQL客户程序与服务器之间的本地通信指定一个套接字文件
lower_case_table_names	此参数不可以动态修改，必须重启数据库。=1：表示存储在磁盘是小写的，但是比较的时候不区分大小写；=0：表示存储为给定的大小，比较是区分大小写的；=2，表示存储为给定的大小写，但是比较的时候是小写的
user	用户名
character-set-server	新数据库或数据表的字符集
symbolic-links	数据库或表是否可以存储在my.cnf中指定datadir之外的分区或目录
log-error	错误日志文件路径
pid-file	指定存放进程ID的文件

图7-10　配置my.cnf

按<ESC>键，然后输入:wq，保存退出。

3）配置MySQL兼容性。

由于MySQL需要用到libtinfo.so.5与libncurses.so.5两个文件，本机已存在libtinfo.so.6与libncurses.so.6文件，故创建链接文件，如图7-11所示。

```
#cd /usr/lib64
#ln -sf libtinfo.so.6 libtinfo.so.5
#ln -sf libncurses.so.6 libncurses.so.5
```

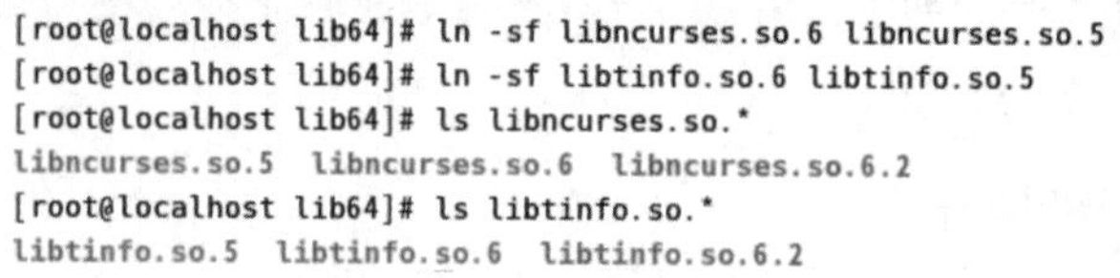

```
[root@localhost lib64]# ln -sf libncurses.so.6 libncurses.so.5
[root@localhost lib64]# ln -sf libtinfo.so.6 libtinfo.so.5
[root@localhost lib64]# ls libncurses.so.*
libncurses.so.5  libncurses.so.6  libncurses.so.6.2
[root@localhost lib64]# ls libtinfo.so.*
libtinfo.so.5  libtinfo.so.6  libtinfo.so.6.2
```

图7-11　创建链接文件

说明

这是 MySQL 5.7.24 与 CentOS Stream 9 系统的兼容性问题，如果 MySQL 安装在其他系统中，可能不存在该问题。

（3）初始化MySQL

初始化数据库，如图7-12所示。

```
# mysqld --initialize-insecure --basedir=/usr/local/mysql --datadir=/usr/local/mysql/data --user=root
```

注意

这是一整行命令，命令中所有路径都是绝对路径。

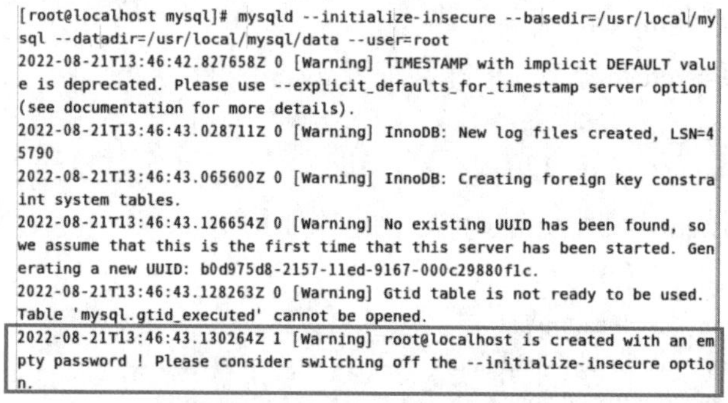

```
[root@localhost mysql]# mysqld --initialize-insecure --basedir=/usr/local/my
sql --datadir=/usr/local/mysql/data --user=root
2022-08-21T13:46:42.827658Z 0 [Warning] TIMESTAMP with implicit DEFAULT valu
e is deprecated. Please use --explicit_defaults_for_timestamp server option
(see documentation for more details).
2022-08-21T13:46:43.028711Z 0 [Warning] InnoDB: New log files created, LSN=4
5790
2022-08-21T13:46:43.065600Z 0 [Warning] InnoDB: Creating foreign key constra
int system tables.
2022-08-21T13:46:43.126654Z 0 [Warning] No existing UUID has been found, so
we assume that this is the first time that this server has been started. Gen
erating a new UUID: b0d975d8-2157-11ed-9167-000c29880f1c.
2022-08-21T13:46:43.128263Z 0 [Warning] Gtid table is not ready to be used.
Table 'mysql.gtid_executed' cannot be opened.
2022-08-21T13:46:43.130264Z 1 [Warning] root@localhost is created with an em
pty password ! Please consider switching off the --initialize-insecure optio
n.
```

注意

初始化提示语中指出了 root 用户进入 MySQL 的密码为空。

图7-12　初始化数据库

（4）启动MySQL

将MySQL加入服务，设置开机自启，并启动MySQL服务，如图7-13所示。

注意

需要提前安装 chkconfig，否则不存在 /etc/init.d 路径。

```
# yum install -y chkconfig
# cp /usr/local/mysql/support-files/mysql.server /etc/init.d/mysql
# chkconfig mysql on
# service mysql start
```

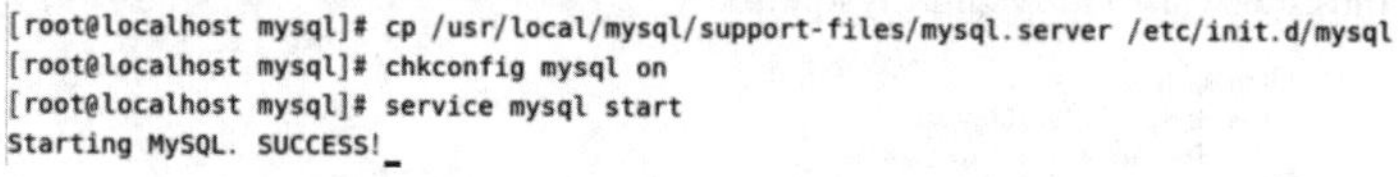

```
[root@localhost mysql]# cp /usr/local/mysql/support-files/mysql.server /etc/init.d/mysql
[root@localhost mysql]# chkconfig mysql on
[root@localhost mysql]# service mysql start
Starting MySQL. SUCCESS!
```

图7-13　设置MySQL开机启动

登录MySQL，密码为空，直接按<Enter>键，如图7-14所示。

```
# mysql -u root -p
```

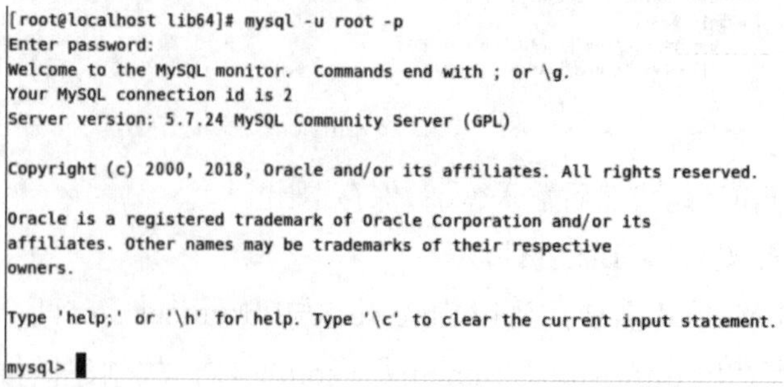

```
[root@localhost lib64]# mysql -u root -p
Enter password:
Welcome to the MySQL monitor.  Commands end with ; or \g.
Your MySQL connection id is 2
Server version: 5.7.24 MySQL Community Server (GPL)

Copyright (c) 2000, 2018, Oracle and/or its affiliates. All rights reserved.

Oracle is a registered trademark of Oracle Corporation and/or its
affiliates. Other names may be trademarks of their respective
owners.

Type 'help;' or '\h' for help. Type '\c' to clear the current input statement.

mysql>
```

图7-14　登录MySQL

设置root用户的密码，如图7-15所示。

> set password='123456';

```
mysql> set password='123456';
Query OK, 0 rows affected (0.01 sec)
```

图7-15　设置密码

> **注意**
>
> 在实际生产过程中，务必为数据库用户设置一个复杂的密码。

（5）配置hive库与hive用户

1）配置hive库。

创建hive库，如图7-16所示。

> create database hive default charset utf8;

```
mysql> create database hive default charset utf8;
Query OK, 1 row affected (0.00 sec)
```

图7-16　创建hive库

> **说明**
>
> 在MySQL中创建Hive数据库，专门用于存储Hive中的元数据信息。

查看数据库，如图7-17所示。

> show databases;

```
mysql> show databases;
+--------------------+
| Database           |
+--------------------+
| information_schema |
| hive               |
| mysql              |
| performance_schema |
| sys                |
+--------------------+
5 rows in set (0.00 sec)
```

图7-17　查看数据库

2）配置hive用户。

创建hive用户，设置密码为hive，并赋予远程登录权限，如图7-18所示。

> create user hive@localhost identified by 'hive';
> grant all on *.* to hive@localhost;
> flush privileges;

```
mysql> create user hive@localhost identified by 'hive';
Query OK, 0 rows affected (0.00 sec)

mysql> grant all on *.* to hive@localhost;
Query OK, 0 rows affected (0.00 sec)

mysql> flush privileges;
Query OK, 0 rows affected (0.00 sec)
```

图7-18　赋予权限

> **说明**
>
> MySQL为了安全性，在默认情况下只允许用户在本地登录，可是在有些情况下，还是需要进行远程连接，因此需要进行赋权。

退出MySQL。

\# exit

4. 配置本地模式Hive

切换为hadoop用户。

\#su - hadoop

扫码看视频

(1)配置hive-env.sh文件

将"/usr/local/hive/conf"目录下的hive-env.sh.template文件重命名为hive-env.sh。

```
$ cd /usr/local/hive/conf
$ cp hive-env.sh.template hive-env.sh
$ vi hive-env.sh
```

将HADOOP_HOME前面的#去掉，使变量生效，并为其指定正确的HADOOP_HOME路径，如图7-19所示。

```
export HADOOP_HOME=/usr/local/hadoop
```

```
# Set HADOOP_HOME to point to a specific hadoop install directory
HADOOP_HOME=/usr/local/hadoop
```

图7-19 修改hive-env.sh文件

(2)配置hive-default.xml文件

将"/usr/local/hive/conf"目录下的hive-default.xml.template文件重命名为hive-default.xml。

```
$ cd /usr/local/hive/conf
$ mv hive-default.xml.template hive-default.xml
```

(3)配置hive-site.xml文件

新建hive-site.xml。

```
$ vi hive-site.xml
```

在hive-site.xml中添加如下内容并保存，如图7-20所示。

```
<?xml version="1.0" encoding="UTF-8" standalone="no"?>
<?xml-stylesheet type="text/xsl" href="configuration.xsl"?>
<configuration>
  <property>
    <name>javax.jdo.option.ConnectionURL</name>
    <value>jdbc:mysql://localhost:3306/hive?createDatabaseIfNotExist=true&amp;useSSL=false</value>
    <description>JDBC connect string for a JDBC metastore</description>
  </property>
  <property>
    <name>javax.jdo.option.ConnectionDriverName</name>
    <value>com.mysql.jdbc.Driver</value>
    <description>Driver class name for a JDBC metastore</description>
  </property>
  <property>
    <name>javax.jdo.option.ConnectionUserName</name>
    <value>hive</value>
    <description>username to use against metastore database</description>
  </property>
  <property>
    <name>javax.jdo.option.ConnectionPassword</name>
    <value>hive</value>
    <description>password to use against metastore database</description>
  </property>
</configuration>
```

⚠ 注意

配置文件不要写错。如果写错，会导致很多莫名其妙的错误，很难排查。

📝 笔记

javax.jdo.option.ConnectionURL：默认使用Hive自带的Derby数据库，这里配置MySQL作为元数据库。

📝 说明

MySQL默认的数据通道是不加密的，在一些安全性要求特别高的场景下，需要配置MySQL端口为SSL，使得数据通道加密处理，避免敏感信息泄露和被篡改。

📝 笔记

javax.jdo.option.ConnectionDriverName：配置数据库连接驱动。

javax.jdo.option.ConnectionUserName：配置连接MySQL的hive用户。

javax.jdo.option.ConnectionPassword：配置连接MySQL的hive用户密码。

```
<?xml version="1.0" encoding="UTF-8" standalone="no"?>
<?xml-stylesheet type="text/xsl" href="configuration.xsl"?>
<configuration>
  <property>
    <name>javax.jdo.option.ConnectionURL</name>
    <value>jdbc:mysql://localhost:3306/hive?createDatabaseIfNotExist=true&am
p;useSSL=false</value>
    <description>JDBC connect string for a JDBC metastore</description>
  </property>
  <property>
    <name>javax.jdo.option.ConnectionDriverName</name>
    <value>com.mysql.jdbc.Driver</value>
    <description>Driver class name for a JDBC metastore</description>
  </property>
  <property>
    <name>javax.jdo.option.ConnectionUserName</name>
    <value>hive</value>
    <description>username to use against metastore database</description>
  </property>
  <property>
    <name>javax.jdo.option.ConnectionPassword</name>
    <value>hive</value>
    <description>password to use against metastore database</description>
  </property>
</configuration>
```

图7-20　hive-site.xml文件

（4）安装MySQL的JDBC驱动包

访问MySQL官网（https://downloads.mysql.com）下载安装文件mysql-connector-java-5.1.47.tar.gz，如图7-21所示。

```
$ cd /usr/local
$sudo wget https://downloads.mysql.com/archives/get/p/3/file/mysql-connector-java-5.1.47.tar.gz
```

```
[hadoop@localhost local]$ sudo wget https://downloads.mysql.com/archives/get/p/3
/file/mysql-connector-java-5.1.47.tar.gz
--2022-08-21 07:33:50--  https://downloads.mysql.com/archives/get/p/3/file/mysql
-connector-java-5.1.47.tar.gz
正在解析主机 downloads.mysql.com (downloads.mysql.com)... 198.18.9.152
正在连接 downloads.mysql.com (downloads.mysql.com)|198.18.9.152|:443... 已连接。
已发出 HTTP 请求，正在等待回应... 302 Moved Temporarily
位置：https://cdn.mysql.com/archives/mysql-connector-java-5.1/mysql-connector-ja
va-5.1.47.tar.gz [跟随至新的 URL]
--2022-08-21 07:33:51--  https://cdn.mysql.com/archives/mysql-connector-java-5.1
/mysql-connector-java-5.1.47.tar.gz
正在解析主机 cdn.mysql.com (cdn.mysql.com)... 198.18.9.153
正在连接 cdn.mysql.com (cdn.mysql.com)|198.18.9.153|:443... 已连接。
已发出 HTTP 请求，正在等待回应... 200 OK
长度：4452049 (4.2M) [application/x-tar-gz]
正在保存至："mysql-connector-java-5.1.47.tar.gz"

mysql-connector-jav 100%[===================>]   4.25M  3.97MB/s  用时 1.1s

2022-08-21 07:33:53 (3.97 MB/s) - 已保存 "mysql-connector-java-5.1.47.tar.gz" [4
452049/4452049])
```

图7-21　下载JDBC安装文件

> ⚠ **注意**
>
> 下载过程需保持网络状态。

解压JDBC安装文件：

```
$ sudo tar -zxf mysql-connector-java-5.1.47.tar.gz
```

将mysql-connector-java-5.1.47-bin.jar复制到/usr/local/hive/lib路径下，如图7-22所示。

```
$ cp mysql-connector-java-5.1.47/mysql-connector-java-5.1.47-bin.jar /usr/local/hive/lib
```

```
[hadoop@localhost local]$ sudo tar -zxf mysql-connector-java-5.1.47.tar.gz
[hadoop@localhost local]$ cp mysql-connector-java-5.1.47/mysql-connector-java-5.
1.47-bin.jar /usr/local/hive/lib/
[hadoop@localhost local]$ ls /usr/local/hive/lib/mysql-*
/usr/local/hive/lib/mysql-connector-java-5.1.47-bin.jar
/usr/local/hive/lib/mysql-metadata-storage-0.12.0.jar
```

图7-22　复制JDBC的jar包

5. 启动本地模式Hive

（1）初始化元数据

初始化元数据，如图7-23所示。

```
$schematool -dbType mysql -initSchema
```

```
[hadoop@localhost local]$ schematool -dbType mysql -initSchema
SLF4J: Class path contains multiple SLF4J bindings.
SLF4J: Found binding in [jar:file:/usr/local/hive/lib/log4j-slf4j-impl-2.10.0.ja
r!/org/slf4j/impl/StaticLoggerBinder.class]
SLF4J: Found binding in [jar:file:/usr/local/hadoop/share/hadoop/common/lib/slf4
j-reload4j-1.7.36.jar!/org/slf4j/impl/StaticLoggerBinder.class]
SLF4J: See http://www.slf4j.org/codes.html#multiple_bindings for an explanation.
SLF4J: Actual binding is of type [org.apache.logging.slf4j.Log4jLoggerFactory]
Metastore connection URL:        jdbc:mysql://localhost:3306/hive?createDatabase
IfNotExist=true&useSSL=false
Metastore Connection Driver :    com.mysql.jdbc.Driver
Metastore connection User:       hive
Starting metastore schema initialization to 3.1.0
Initialization script hive-schema-3.1.0.mysql.sql
```

图7-23 初始化元数据

（2）启动Hadoop

启动Hive前，先启动Hadoop，并看到NameNode、DataNode、SecondaryNameNode、NodeManager、ResourceManager进程。

```
$ start-all.sh
```

（3）启动Hive

启动Hive，如图7-24所示。

```
$hive
```

```
[hadoop@localhost local]$ hive
SLF4J: Class path contains multiple SLF4J bindings.
SLF4J: Found binding in [jar:file:/usr/local/hbase/lib/client-facing-thirdparty/
slf4j-reload4j-1.7.33.jar!/org/slf4j/impl/StaticLoggerBinder.class]
SLF4J: Found binding in [jar:file:/usr/local/hive/lib/log4j-slf4j-impl-2.10.0.ja
r!/org/slf4j/impl/StaticLoggerBinder.class]
SLF4J: Found binding in [jar:file:/usr/local/hadoop/share/hadoop/common/lib/slf4
j-reload4j-1.7.36.jar!/org/slf4j/impl/StaticLoggerBinder.class]
SLF4J: See http://www.slf4j.org/codes.html#multiple_bindings for an explanation.
SLF4J: Actual binding is of type [org.slf4j.impl.Reload4jLoggerFactory]
2022-08-21 07:41:27,402 INFO  [main] conf.HiveConf: Found configuration file fil
e:/usr/local/hive/conf/hive-site.xml
Hive Session ID = c58d135e-2ec9-456c-9700-cc8880471170
```

图7-24 启动Hive

退出Hive。

```
>exit;
```

Hive已经正常启动，但使用过程中会出现非常多的info提示。可以通过配置conf路径下log4j.properties日志文件取消显示info提示。

```
$ vi log4j.properties
```

编辑内容如下：

```
log4j.rootLogger= CA
log4j.appender.CA=org.apache.log4j.ConsoleAppender
log4j.appender.CA.layout=org.apache.log4j.PatternLayout
log4j.appender.CA.layout.ConversionPattern=%-4r
[%t] %-5p %c %x - %m%n
```

再次启动Hive，将不会有info提示。

⚠️ 注意

由于Hive需要使用Hadoop中的MapReduce与HDFS，可以认为Hive是部署在Hadoop集群之上的。因此，需要保证在Hive运行的整个生命周期中Hadoop都是正常运行的。

启动与关闭Hive的顺序是：启动Hadoop→启动Hive→关闭Hive→关闭Hadoop。

项目 7　部署与使用 Hive

扫码看视频

任务2　部署远程模式Hive

任务描述

王工在Linux系统中成功部署本地模式Hive后，发现在本地模式中Hive客户端和Metastore服务只能在同一台服务器中启动，不能实现多节点并发调用Hive的需求，希望在Linux系统中部署远程模式Hive。

远程模式分为客户端与服务端两个部分。服务端的配置与本地模式相同，客户端需要单独配置。本次任务需要在Hadoop完全分布环境中进行，将master节点作为Hive的服务端，slave1节点作为Hive的客户端。

任务分析

1. 任务目标

1）理解Hive的工作原理及体系架构。

2）掌握Hive单机部署。

3）能够简单运行Hive程序。

2. 任务环境

操作系统：CentOS Stream 9（预装分布式Hadoop）

软件版本：Java 1.8.0、MySQL 5.7.24、Hadoop3.3.4、Hive 3.1.2

3. 任务导图

任务导图如图7-25所示。

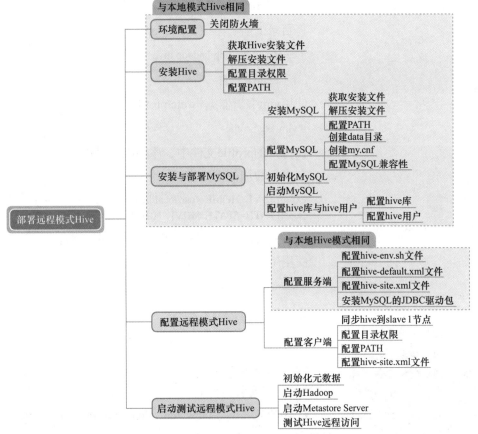

图7-25　任务导图

135

> **说明**
> 在远程模式 Hive 中，客户端需要远程访问服务端，因此关闭防火墙。

> **说明**
> 安装 Hive 过程与任务 1 相同，详细内容参照任务 1，此处仅罗列必要步骤。

> **说明**
> 安装与部署 MySQL 过程与任务 1 相同，详细内容参照任务 1，此处仅罗列必要步骤。

任务实施

1. 环境配置

关闭防火墙

运行指令，关闭所有节点的防火墙，并查看防火墙状态，确认防火墙已关闭。

```
# systemctl stop firewalld
# systemctl status firewalld
```

2. 安装Hive

在master节点完成以下部署。

（1）获取Hive安装文件

访问Hive官网（https://hive.apache.org/）下载安装文件apache-hive-3.1.2-bin.tar.gz。

```
# cd /usr/local
# wget https://downloads.apache.org/hive/hive-3.1.2/apache-hive-3.1.2-bin.tar.gz
```

（2）解压安装文件

```
# tar -zxf apache-hive-3.1.2-bin.tar.gz
```

更改目录名称为hive。

```
# mv apache-hive-3.1.2-bin hive
```

（3）配置目录权限

将目录的所有者改为hadoop用户。

```
# chown -R hadoop:hadoop hive
```

（4）配置PATH

使用vim编辑器编辑/etc/profile。

```
# vim /etc/profile
```

在打开的profile文件中，添加Hive相关路径。

在末行增加如下内容：

```
export HIVE_HOME=/usr/local/hive
export PATH=$PATH:$HIVE_HOME/bin
```

配置完成后，按<ESC>键，输入:wq，保存并退出。

使用source指令，使新配置生效。

```
# source /etc/profile
```

3. 安装与部署MySQL

在master节点完成以下部署。

（1）安装MySQL

1）获取安装文件。

从MySQL官网（https://downloads.mysql.com/archives/community/）下载安装文件mysql-*.*.*.tar.gz，例如mysql-5.7.24-linux-glibc2.12

-x86_64.tar.gz：

```
# cd /usr/local
# wget https://downloads.mysql.com/archives/get/p/23/file/mysql-5.7.24-linux-glibc2.12-x86_64.tar.gz
```

2）解压安装文件。

解压MySQL安装包，并重命名为mysql。

```
# tar -zxf mysql-5.7.24-linux-glibc2.12-x86_64.tar.gz
# mv mysql-5.7.24-linux-glibc2.12-x86_64 mysql
```

3）配置PATH。

配置MySQL安装路径至PATH变量。

```
# vim /etc/profile
```

添加如下内容：

```
export MYSQL_HOME=/usr/local/mysql
export PATH=$PATH:$MYSQL_HOME/bin
```

按<ESC>键，然后输入:wq，保存并退出。

使用source命令使得环境变量生效。

```
# source /etc/profile
```

（2）配置MySQL

1）创建data目录。

进入mysql路径创建data目录。

```
# cd /usr/local/mysql
# mkdir data
```

注意

未创建 data 目录，可能会导致初始化失败。

2）创建my.cnf。

在系统/etc路径下，创建并配置my.cnf文件。

```
# cd /etc
# vi my.cnf
```

在my.cnf中添加如下内容。

```
[mysqld]
basedir=/usr/local/mysql
datadir=/usr/local/mysql/data
socket=/tmp/mysql.sock
lower_case_table_names=1
user=root
#default-character-set=utf8
character-set-server=utf8
# Disabling symbolic-links is recommended to prevent assorted security risks
symbolic-links=0
[client]
default-character-set=utf8
[mysqld_safe]
log-error=/var/log/mysqld.log
pid-file=/var/run/mysqld/mysqld.pid
```

按<ESC>键，然后输入:wq，保存并退出。

3）配置MySQL兼容性。

由于MySQL需要用到libtinfo.so.5与libncurses.so.5两个文件，本

说明

这是 MySQL 5.7.24 与 CentOS Stream 9 系统的兼容性问题，如果 MySQL 安装在其他系统中，可能不存在该问题。

机已存在libtinfo.so.6与libncurses.so.6文件，故创建链接文件。

```
#cd /usr/lib64
#ln -sf libtinfo.so.6 libtinfo.so.5
#ln -sf libncurses.so.6 libncurses.so.5
```

⚠️ **注意**

这是一整行命令，命令中所有路径都是绝对路径。

（3）初始化MySQL

初始化数据库。

```
# mysqld --initialize-insecure --basedir=/usr/local/mysql --datadir=/usr/local/mysql/data --user=root
```

（4）启动MySQL

将MySQL加入服务，设置开机自启，并启动MySQL服务。

```
# yum install -y chkconfig
# cp /usr/local/mysql/support-files/mysql.server /etc/init.d/mysql
# chkconfig mysql on
# service mysql start
```

登录MySQL，密码为空，直接按<Enter>键。

```
# mysql -u root -p
```

设置root用户的密码。

```
> set password='123456';
```

（5）配置hive库与hive用户

1）配置hive库。

创建hive库。

```
> create database hive default charset utf8;
```

查看数据库。

```
> show databases;
```

2）配置hive用户。

📝 **说明**

MySQL 为了安全性，在默认情况下用户只允许在本地登录，可是在有些情况下，还是需要进行远程连接，因此需要进行赋权。

创建hive用户，设置密码为hive，并赋予远程登录权限。

```
> create user hive@localhost identified by 'hive';
> grant all on *.* to hive@localhost;
> flush privileges;
```

退出MySQL。

```
# exit
```

4. 配置远程模式Hive

（1）配置服务端

本任务将master作为服务端，因此在master节点完成以下部署。

切换为hadoop用户。

```
#su - hadoop
```

📝 **说明**

配置远程模式 Hive 的服务器端与任务 1 相同，详细内容参照任务 1，此处仅罗列必要步骤。

1）配置hive-env.sh文件。

将"/usr/local/hive/conf"目录下的hive-env.sh.template文件重命名为hive-env.sh。

```
$ cd /usr/local/hive/conf
$ cp hive-env.sh.template hive-env.sh
$ vi hive-env.sh
```

将HADOOP_HOME前面的#去掉，使变量生效，并为其指定正确的HADOOP_HOME路径。

```
export HADOOP_HOME=/usr/local/hadoop
```

2）配置hive-default.xml文件。

将"/usr/local/hive/conf"目录下的hive-default.xml.template文件重命名为hive-default.xml。

```
$ cd /usr/local/hive/conf
$ mv hive-default.xml.template hive-default.xml
```

3）配置hive-site.xml文件。

新建hive-site.xml。

```
$ vi hive-site.xml
```

在hive-site.xml中添加如下内容并保存：

```
<?xml version="1.0" encoding="UTF-8" standalone="no"?>
<?xml-stylesheet type="text/xsl" href="configuration.xsl"?>
<configuration>
  <property>
    <name>javax.jdo.option.ConnectionURL</name>
    <value>jdbc:mysql://localhost:3306/hive?createDatabaseIfNotExist=true&amp;useSSL=false</value>
    <description>JDBC connect string for a JDBC metastore</description>
  </property>
  <property>
    <name>javax.jdo.option.ConnectionDriverName</name>
    <value>com.mysql.jdbc.Driver</value>
    <description>Driver class name for a JDBC metastore</description>
  </property>
  <property>
    <name>javax.jdo.option.ConnectionUserName</name>
    <value>hive</value>
    <description>username to use against metastore database</description>
  </property>
  <property>
    <name>javax.jdo.option.ConnectionPassword</name>
    <value>hive</value>
    <description>password to use against metastore database</description>
  </property>
</configuration>
```

4）安装MySQL的JDBC驱动包。

访问MySQL官网（https://downloads.mysql.com）下载安装文件mysql-connector-java-5.1.47.tar.gz。

```
$ cd /usr/local
$ sudo wget https://downloads.mysql.com/archives/get/p/3/file/mysql-connector-java-5.1.47.tar.gz
```

📝 **笔记**

javax.jdo.option.ConnectionURL：默认使用Hive自带的Derby数据库，这里配置MySQL作为元数据库。

📝 **说明**

MySQL默认的数据通道是不加密的，在一些安全性要求特别高的场景下，需要配置MySQL端口为SSL，使得数据通道加密处理，避免敏感信息泄露和被篡改。

📝 **笔记**

javax.jdo.option.ConnectionDriverName：配置数据库连接驱动。

javax.jdo.option.ConnectionUserName：配置连接MySQL的hive用户。

javax.jdo.option.ConnectionPassword：配置连接MySQL的hive用户密码。

⚠️ **注意**

下载过程需保持网络状态。

解压JDBC安装文件。

```
$ sudo tar -zxf mysql-connector-java-5.1.47.tar.gz
```

将mysql-connector-java-5.1.47-bin.jar复制到/usr/local/hive/lib路径下。

```
$ cp mysql-connector-java-5.1.47/mysql-connector-java-5.1.47-bin.jar /usr/local/hive/lib
```

（2）配置客户端

1）同步hive到slave1节点。

在master节点，复制hive文件夹至slave1节点，如图7-26所示。

```
# scp -r /usr/local/hive/ root@slave1:/usr/local/
```

```
[hadoop@master local]$ scp -r /usr/local/hive/ root@slave1:/usr/local
root@slave1's password:
LICENSE                                        100%   20KB  10.8MB/s   00:00
NOTICE                                         100%  230   261.6KB/s   00:00
RELEASE_NOTES.txt                              100% 2469    1.4MB/s    00:00
asm-LICENSE                                    100% 1511    1.4MB/s    00:00
com.google.protobuf-LICENSE                    100% 2133    1.9MB/s    00:00
```

图7-26　同步hive到slave1节点

2）配置目录权限。

在slave1节点，修改hive文件夹的所有者为hadoop用户，如图7-27所示。

```
# cd /usr/local
# chown -R hadoop:hadoop hive
```

```
[root@slave1 ~]# cd /usr/local
[root@slave1 local]# ls
bin    games   hbase    include   lib64    sbin   src
etc    hadoop  hive     lib       libexec  share
[root@slave1 local]# chown -R hadoop:hadoop /usr/local/hive
[root@slave1 local]# ls -ld hive/
drwxr-xr-x. 10 hadoop_hadoop 184 11月  1 03:49 hive/
```

图7-27　修改slave1节点的hive文件夹所有者

3）配置PATH。

在slave1节点，使用vim编辑器编辑/etc/profile。

```
# vim /etc/profile
```

在末行增加如下内容：

```
export HIVE_HOME=/usr/local/hive
export PATH=$PATH:$HIVE_HOME/bin
```

保存并退出后，运行指令使配置生效。

```
# source /etc/profile
```

4）配置hive-site.xml文件。

在slave1节点，编辑/usr/local/hive/conf下的hive-site.xml。

```
# su - hadoop
$ vi /usr/local/hive/conf/hive-site.xml
```

在hive-site.xml中添加如下内容，如图7-28所示。

```xml
<?xml version="1.0" encoding="UTF-8" standalone="no"?>
<?xml-stylesheet type="text/xsl" href="configuration.xsl"?>
<configuration>
 <property>
  <name>hive.metastore.local</name>
  <value>false</value>
 </property>
 <property>
  <name>hive.metastore.uris</name>
  <value>thrift://master:9083</value>
 </property>
</configuration>
```

> **笔记**
>
> **hive-site.xml 内容解析：**
>
> hive.metastore.local：是否启用本地服务器连接hive，false 为非本地模式，即远程模式。
>
> hive.metastore.uris：hive服务端Metastore Server连接地址，默认监听端口为9083。

```xml
<?xml version="1.0" encoding="UTF-8" standalone="no"?>
<?xml-stylesheet type="text/xsl" href="configuration.xsl"?>
<configuration>
  <property>
    <name>hive.metastore.local</name>
    <value>false</value>
  </property>
  <property>
    <name>hive.metastore.uris</name>
    <value>thrift://master:9083</value>
  </property>
</configuration>
```

图7-28　slave1节点的hive-site.xml配置

5. 启动测试远程模式Hive

（1）服务端初始化元数据

`$schematool -dbType mysql -initSchema`

（2）启动Hadoop

启动Hive前，先启动Hadoop，在master节点执行指令并看到NameNode、DataNode、SecondaryNameNode、NodeManager、ResourceManager进程。

`$ start-all.sh`

（3）启动Metastore Server

如图7-29所示，在master节点启动Metastore Server。

`$hive --service metastore`

```
[hadoop@master hive]$ hive --service metastore
2022-09-20 18:17:52: Starting Hive Metastore Server
SLF4J: Class path contains multiple SLF4J bindings.
SLF4J: Found binding in [jar:file:/usr/local/hive/lib/log4j-slf4j-impl-
2.10.0.jar!/org/slf4j/impl/StaticLoggerBinder.class]
SLF4J: Found binding in [jar:file:/usr/local/hadoop/share/hadoop/common
/lib/slf4j-reload4j-1.7.36.jar!/org/slf4j/impl/StaticLoggerBinder.class
]
SLF4J: See http://www.slf4j.org/codes.html#multiple_bindings for an exp
lanation.
SLF4J: Actual binding is of type [org.apache.logging.slf4j.Log4jLoggerF
actory]
```

图7-29　启动Metastore Server

此时，通过jps命令在master节点查看进程，会发现多了一个名为"RunJar"的进程，如图7-30所示，它就是Metastore Server的独立进程。

> **⚠ 注意**
>
> 由于Hive需要使用Hadoop中的MapReduce与HDFS，可以认为Hive是部署在Hadoop集群之上的。因此，需要保证在Hive运行的整个生命周期中Hadoop都是正常运行的。
>
> 启动与关闭Hive的顺序是：启动Hadoop→启动Hive→关闭Hive→关闭Hadoop。

> **说明**
>
> Hive的Web界面功能从Hive 2.2.0版本起被删除了。在Hive 2.2.0版本前，可以通过10002端口访问。

```
[hadoop@master local]$ jps
7908 NameNode
8244 SecondaryNameNode
10006 RunJar
10247 Jps
10170 RunJar
```

图7-30　查看进程

在master中启动Hive。

```
$ hive
```

使用jps查看Java进程，则会再次产生一个"RunJar"进程，如图7-31所示，该进程即为Hive的远程模式中的客户进程。

```
[hadoop@master local]$ jps
7908 NameNode
8244 SecondaryNameNode
10006 RunJar
10247 Jps
10170 RunJar
```

图7-31　查看Hive进程

（4）测试Hive远程访问

在slave1节点启动Hive。

```
$ hive
```

这样就在master和slave1两个节点中分别启动了Hive的客户进程。

接下来，验证在master节点上的操作是否能在slave1节点中查看到效果。

在master节点的Hive中创建表student（默认建在default数据库中），如图7-32所示。

```
>create table student(id int, name string);
```

```
hive> create table student(id int,name string);
OK
Time taken: 6.716 seconds
hive>
```

图7-32　在master节点的Hive中创建表student

在slave1节点的Hive客户端中查询所有表，可以看到master节点创建的表student，如图7-33所示，说明Hive远程模式配置成功。

```
>show tables;
```

```
hive> show tables;
OK
student
Time taken: 1.166 seconds, Fetched: 1 row(s)
```

图7-33　在slave1节点的Hive中成功查询表student

项目 7　部署与使用 Hive

扫码看视频

任务3　利用Hive实现数据导入

任务描述

王工为客户成功部署Hive后，客户希望将现有文本文件中的内容批量导入到Hive数据仓库中。现有文本文件内容如下，四列内容依次表示：ID、姓名、年龄、是否为男性。

```
1 zhangsan 15 true
2 lisi 25 false
```

此外，再手动插入1条员工信息：ID为3，姓名为wangwu，年龄18，男性。

最终在Hive中生成数据表，见表7-5。

表7-5　人员信息表

ID	姓　名	年　　龄	是 否 男 性
1	zhangsan	15	true
2	lisi	25	false
3	wangwu	18	true

任务分析

1. 任务目标

1）掌握Hive的基本操作。

2）熟悉Hive的基本语法。

2. 任务环境

操作系统：CentOS Stream 9（预装分布式Hadoop+Hive）

软件版本：Java 1.8.0、MySQL 5.7.24、Hadoop3.3.4、Hive 3.1.2

3. 任务导图

任务导图如图7-34所示。

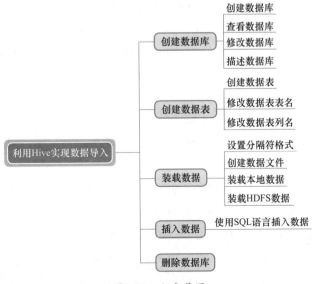

图7-34　任务导图

143

任务实施

1. 创建数据库

（1）创建数据库

创建数据库，如图7-35所示，名称为hive。

```
hive> create database hive;
```

```
hive> create database hive;
OK
Time taken: 0.032 seconds
```

图7-35 创建数据库hive

> **说明**
>
> 本任务实施前需要启动 Hadoop 与 Hive。

> **笔记**
>
> 创建数据库hive，如果同名数据库已经存在，则会抛出异常，加上if not exists关键字，则不会抛出异常，如图7-38所示。
>
> hive> create database if not exists hive;
>
> ```
> hive> create database if not exists hive;
> OK
> Time taken: 0.245 seconds
> ```
>
> 图7-38 加上关键字创建数据库

（2）查看数据库

查看Hive中包含的所有数据库，如图7-36所示。

```
hive> show databases;
```

```
hive> show databases;
OK
default
hive
Time taken: 0.139 seconds, Fetched: 2 row(s)
```

图7-36 查看数据库

查看Hive中以h开头的所有数据库，如图7-37所示。

```
hive> show databases like 'h*';
```

```
hive> show databases like 'h*';
OK
hive
Time taken: 0.045 seconds, Fetched: 1 row(s)
```

图7-37 查看以h开头的数据库

（3）修改数据库

为hive数据库设置dbproperties键值对属性值来描述数据库属性信息，如图7-39所示。

```
hive> alter database hive set dbproperties ('owner'='hadoop');
```

```
hive> describe database extended hive;
OK
hive    hdfs://localhost:9000/user/hive/warehouse/hive.db    hadoop  USER
 {owner=hadoop}
Time taken: 0.03 seconds, Fetched: 1 row(s)
```

图7-39 设置dbproperties键值对属性值

（4）描述数据库

查看数据库hive的基本信息，包括数据库文件位置信息等，如图7-40所示。

```
hive> describe database hive;
```

```
hive> describe database hive;
OK
hive    hdfs://localhost:9000/user/hive/warehouse/hive.db    hadoop USER
Time taken: 0.039 seconds, Fetched: 1 row(s)
```

图7-40 查看数据库hive的基本信息

> **说明**
>
> 从反馈结果可知，hive数据库的文件存放在HDFS文件系统中，路径为 /user/hive/warehouse。

查看数据库hive的详细信息，包括数据库的基本信息及属性信息等，如图7-41所示。

```
hive>describe database extended hive;
```

```
hive> describe database extended hive;
OK
hive         hdfs://localhost:9000/user/hive/warehouse/hive.db        hadoop    USER
{owner=hadoop}
Time taken: 0.03 seconds, Fetched: 1 row(s)
```

图7-41　查看数据库hive的详细信息

> **说明**
> 从反馈结果可知，hive数据库有一个属性owner，属性值为hadoop，是在（3）中添加的。

2. 创建数据表

（1）创建数据表

在hive数据库中，创建内部表usr，包含三个属性：id，name，age，存储路径为"/usr/local/hive/warehouse/hive/usr"，如图7-42所示。

```
hive>create table if not exists hive.usr(id bigint,name string,age int) location '/usr/local/hive/warehouse/hive/usr';
```

```
hive> create table if not exists usr(id bigint,name string,age int)
    > location '/usr/local/hive/warehouse/hive/usr';
OK
Time taken: 0.128 seconds
```

图7-42　创建数据内部表usr

> **笔记**
> 内部表：也叫托管表，是Hive在创建表时的默认表。内部表是Hive默认表类型，表数据默认存储在warehouse目录中。在加载数据的过程中，实际数据会被移动到warehouse目录中，当删除表时，表的数据和元数据将会被同时删除。
> 适用于：①ETL数据清理时，使用内部表作为中间表，HDFS上的文件同步删除；②在误删的情况下，易于恢复的数据。

查看数据库hive中所有的数据表和视图，如图7-43所示。

```
hive> use hive;
hive> show tables;
```

```
hive> use hive;
OK
Time taken: 0.02 seconds
hive> show tables;
OK
usr
Time taken: 0.032 seconds, Fetched: 1 row(s)
```

图7-43　查看所有的数据表和视图

> **笔记**
> 查看数据表也可以使用通配符，例如，查看hive数据库中所有以u开头的所有数据表和视图：show tables in hive like 'u*';

查看数据表usr的基本信息，包括列信息，如图7-44所示。

```
hive> describe usr;
```

```
hive> describe usr;
OK
id                      bigint
name                    string
age                     int
Time taken: 0.046 seconds, Fetched: 3 row(s)
```

图7-44　查看数据表usr的基本信息

查看表usr的详细信息，包括列信息、位置信息、属性信息等，如图7-45所示。

```
hive> describe extended usr;
```

```
hive> describe extended usr;
OK
id                      bigint
name                    string
age                     int

Detailed Table Information      Table(tableName:usr, dbName:hive, owner:hado
op, createTime:1667312767, lastAccessTime:0, retention:0, sd:StorageDescript
or(cols:[FieldSchema(name:id, type:bigint, comment:null), FieldSchema(name:n
ame, type:string, comment:null), FieldSchema(name:age, type:int, comment:nul
l)], location:hdfs://master:9000/usr/local/hive/warehouse/hive/usr, inputFor
mat:org.apache.hadoop.mapred.TextInputFormat, outputFormat:org.apache.hadoop
.hive.ql.io.HiveIgnoreKeyTextOutputFormat, compressed:false, numBuckets:-1,
serdeInfo:SerDeInfo(name:null, serializationLib:org.apache.hadoop.hive.serde
2.lazy.LazySimpleSerDe, parameters:{serialization.format=1}), bucketCols:[],
 sortCols:[], parameters:{}, skewedInfo:SkewedInfo(skewedColNames:[], skewed
ColValues:[], skewedColValueLocationMaps:{}), storedAsSubDirectories:false),
 partitionKeys:[], parameters:{totalSize=0, numRows=0, rawDataSize=0, COLUMN
_STATS_ACCURATE={\"BASIC_STATS\":\"true\",\"COLUMN_STATS\":{\"age\":\"true\"
,\"id\":\"true\",\"name\":\"true\"}}, numFiles=0, transient_lastDdlTime=1667
312767, bucketing_version=2}, viewOriginalText:null, viewExpandedText:null,
tableType:MANAGED_TABLE, rewriteEnabled:false, catName:hive, ownerType:USER)
Time taken: 0.119 seconds, Fetched: 5 row(s)
```

图7-45　查看数据表usr的详细信息

> 📝 **笔记**
>
> tableType：MANAGED_TABLE 指表类型为内部表。

（2）修改数据表表名

重命名数据表usr为usr1，如图7-46所示。

```
hive> alter table usr rename to usr1;
```

```
hive> alter table usr rename to usr1;
OK
Time taken: 0.127 seconds
hive> show tables;
OK
usr1
Time taken: 0.031 seconds, Fetched: 1 row(s)
hive>
```

图7-46　重命名数据表usr

（3）修改数据表列名

把数据表usr1中列名name修改为username，该列位于id列后，如图7-47所示。

```
hive>alter table usr1 change name username string after id;
```

```
hive> describe usr1;
OK
id                      bigint
name                    string
age                     int
Time taken: 0.054 seconds, Fetched: 3 row(s)
hive> alter table usr1 change name username string after id;
OK
Time taken: 0.112 seconds
hive> describe usr1;
OK
id                      bigint
username                string
age                     int
Time taken: 0.048 seconds, Fetched: 3 row(s)
```

图7-47　修改数据表usr1中name列为username

在数据表usr1中增加一个新列sex,如图7-48所示。

hive>alter table usr1 add columns(sex boolean);

```
hive> describe usr1;
OK
id                      bigint
username                string
age                     int
Time taken: 0.048 seconds, Fetched: 3 row(s)
hive> alter table usr1 add columns(sex boolean);
OK
Time taken: 0.107 seconds
hive> describe usr1;
OK
id                      bigint
username                string
age                     int
sex                     boolean
Time taken: 0.061 seconds, Fetched: 4 row(s)
```

图7-48　为数据表usr1增加新列sex

3. 装载数据

（1）设置分隔符格式

设置数据表的分隔符，例如，设置为空格，如图7-49所示。

hive>alter table usr1 set SERDEPROPERTIES('field.delim'=' ');

```
hive> describe extended usr1;
OK
id                      bigint
username                string
age                     int
sex                     boolean

Detailed Table Information    Table(tableName:usr1, dbName:hive, owner:had
oop, createTime:1667312767, lastAccessTime:0, retention:0, sd:StorageDescrip
tor(cols:[FieldSchema(name:id, type:bigint, comment:null), FieldSchema(name:
username, type:string, comment:null), FieldSchema(name:age, type:int, commen
t:null), FieldSchema(name:sex, type:boolean, comment:null)], location:hdfs:/
/master:9000/usr/local/hive/warehouse/hive/usr, inputFormat:org.apache.hadoo
p.mapred.TextInputFormat, outputFormat:org.apache.hadoop.hive.ql.io.HiveIgno
reKeyTextOutputFormat, compressed:false, numBuckets:-1, serdeInfo:SerDeInfo(
name:null, serializationLib:org.apache.hadoop.hive.serde2.lazy.LazySimpleSer
De, parameters:{serialization.format=1, field.delim= }), bucketCols:[], sort
Cols:[], parameters:{}, skewedInfo:SkewedInfo(skewedColNames:[], skewedColVa
lues:[], skewedColValueLocationMaps:{}), storedAsSubDirectories:false), part
itionKeys:[], parameters:{last_modified_time=1667313361, totalSize=0, numRow
s=0, rawDataSize=0, COLUMN_STATS_ACCURATE={\"BASIC_STATS\":\"true\",\"COLUMN
_STATS\":{\"age\":\"true\",\"id\":\"true\",\"name\":\"true\"}}, numFiles=0,
transient_lastDdlTime=1667313361, bucketing_version=2, last_modified_by=hado
op}, viewOriginalText:null, viewExpandedText:null, tableType:MANAGED_TABLE,
rewriteEnabled:false, catName:hive, ownerType:USER)
Time taken: 0.071 seconds, Fetched: 6 row(s)
```

图7-49　为数据表usr1设置分隔符

📝 **说明**

Hive中默认的分隔符为^A，并不常用，因此在批量装载数据前，按照导入的数据需求设置数据表分隔符。

如果要在数据文件中输入^A，使用快捷键<Ctrl+V>，再用快捷键<Ctrl+A>。

📝 **说明**

可以看到usr1表的分隔符为空格，另外，可以看到usr1表的数据存放在HDFS文件系统中的位置，后续任务将用到。

（2）创建数据文件

在Linux系统中，创建预装载数据的文件，例如创建usr1.txt文件，如图7-50所示。

```
$ cd /usr/local/hive
$ mkdir data
$ vi data/usr1.txt
```

按照usr1表的各字段数据类型要求，编辑内容：

1 zhangsan 15 true
2 lisi 25 false

```
[hadoop@master ~]$ cd /usr/local/hive
[hadoop@master hive]$ mkdir data
[hadoop@master hive]$ vi data/usr1.txt
[hadoop@master hive]$ cat data/usr1.txt
1    zhangsan  15    true
2    lisi      25    false
```

图7-50 创建预装载数据文件

（3）装载本地数据

把目录"/usr/local/hive/data/usr1.txt"下的数据装载进数据表usr1并覆盖原有数据，查看表内数据，如图7-51所示。

```
hive> load data local inpath '/usr/local/hive/data/usr1.txt' overwrite into table usr1;
hive> select * from usr1;
```

```
hive> load data local inpath '/usr/local/hive/data/usr1.txt' overwrite into table usr1;
Loading data to table hive.usr1
OK
Time taken: 0.25 seconds
hive> select * from usr1;
OK
1       zhangsan        15      true
2       lisi            25      false
Time taken: 0.13 seconds, Fetched: 2 row(s)
```

图7-51 向数据表usr1装载数据

（4）装载HDFS数据

将数据文件上传到HDFS文件系统，如图7-52所示。

```
$ hdfs dfs -mkdir -p /usr/data
$ hdfs dfs -put /usr/local/hive/data/usr1.txt /usr/data
```

```
[hadoop@localhost data]$ hdfs dfs -mkdir -p /usr/data
[hadoop@localhost data]$ hdfs dfs -put /usr/local/hive/data/usr1.txt /usr/data
[hadoop@localhost data]$ hdfs dfs -ls /usr/data
Found 1 items
-rw-r--r--   1 hadoop supergroup         35 2022-08-22 07:32 /usr/data/usr1.txt
```

图7-52 将数据文件上传至HDFS

清空usr1表内的数据，并把分布式文件系统目录"hdfs://localhost:9000/usr/data/usr1.txt"中的数据文件数据装载进usr1表，如图7-53所示。

```
hive> truncate table usr1;
hive> load data inpath '/usr/data/usr1.txt' into table usr1;
```

```
hive> truncate table usr1;
OK
Time taken: 0.18 seconds
hive> select * from usr1;
OK
Time taken: 0.221 seconds
hive> load data inpath '/usr/data/usr1.txt' into table usr1;
Loading data to table hive.usr1
OK
Time taken: 0.181 seconds
hive> select * from usr1;
OK
1       zhangsan        15      true
2       lisi            25      false
Time taken: 0.177 seconds, Fetched: 2 row(s)
```

图7-53　从HDFS向数据表usr1装载数据

从图7-49中可以看到usr1表的数据存放在HDFS的/usr/local/hive/warehouse/hive/usr路径下。

数据装载完成后，查看HDFS中/usr/local/hive/warehouse/hive/usr路径下的文件与HDFS中/usr/data路径（usr1.txt文件原存放路径）下的文件，如图7-54所示。

```
$hdfs dfs -ls /usr/local/hive/warehouse/hive/usr
$hdfs dfs -ls /usr/data
```

```
[hadoop@master hive]$ hdfs dfs -ls /usr/local/hive/warehouse/hive/usr
Found 1 items
-rw-r--r--   3 hadoop supergroup         35 2022-11-01 07:49 /usr/local/hive
/warehouse/hive/usr1.txt
[hadoop@master hive]$ hdfs dfs -ls /usr/data
```

图7-54　usr1.txt文件被移动到新路径下

usr1.txt文件从原来的/usr/data路径移动到了usr1表存放数据的路径，这便是内部表的数据装载过程。

4. 插入数据

为usr1表插入新数据，使用SQL语言，如图7-55所示。

```
hive>insert into usr1 values(3,'wangwu',18,true);
hive>select * from usr1;
```

说明

可以看到，Hive 将 SQL 语句转换成 MapReduce 任务并执行。

说明

最终的结果被移到 HDFS 上。

图7-55 插入数据

再次查看HDFS文件系统中的文件，可以看到新增文件，查看新增文件内容，文件存储了新增数据，如图7-56所示。

```
$hdfs dfs -ls /usr/local/hive/warehouse/hive/usr
$hdfs dfs -cat /usr/local/hive/warehouse/hive/usr/000000_0
```

图7-56 查看HDFS下新增文件内容

5. 删除数据库

如图7-57所示，删除数据库hive，加上cascade关键字，可以删除当前数据库和该数据库中的表。

```
hive> drop database if exists hive cascade;
```

图7-57 删除数据库hive

笔记

删除数据库时，如果库中有数据表，则应加上关键词 cascade，否则系统将会报错。如果不加关键词 cascade，只能删除空数据库，即数据库中无数据表的空数据库。

任务4　利用Hive实现词频统计

扫码看视频

任务描述

王工利用Hive为客户实现词频统计功能,让客户直观地感受到Hive组件相对于MapReduce编程更便捷、更高效。

任务分析

1. 任务目标

1)熟悉Hive的基本使用流程。

2)掌握使用Hive实现简单案例。

2. 任务环境

操作系统:CentOS Stream 9(预装分布式Hadoop+Hive)

软件版本:Java 1.8.0、MySQL 5.7.24、Hadoop3.3.4、Hive 3.1.2

3. 任务导图

任务导图如图7-58所示。

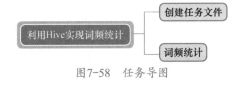

图7-58　任务导图

任务实施

1. 创建任务文件

在实现词频统计之前先设置好需要统计的数据集,并上传到HDFS。

⚠ 注意

任务前,需要正常启动Hadoop。

创建使用的文件word.txt:

```
$ cd /usr/local/hive/data
$ vi word.txt
```

本次实训所使用的word.txt内容为:

```
I love China
China is a great and beautiful country
I am Chinese
I am proud of being Chinese
```

将创建好的文件上传至HDFS,如图7-59所示。

```
$ hdfs dfs -put /usr/local/hive/data/word.txt /usr/data
```

```
[hadoop@localhost data]$ vi word.txt
[hadoop@localhost data]$ hdfs dfs -put /usr/local/hive/data/word.txt /usr/data
[hadoop@localhost data]$ hdfs dfs -cat /usr/data/word.txt
I love China
China is a great and beautiful country
I am Chinese
I am proud of being Chinese
```

图7-59 上传任务数据至HDFS

2. 词频统计

编写HiveQL语句实现WordCount算法。

进入hive命令行界面，编写HiveQL语句实现WordCount算法。

（1）装载统计数据集

首先创建数据表，装载统计数据集，如图7-60所示。

```
hive> create table docs(line string);
hive> load data inpath '/usr/data/word.txt' overwrite into table docs;
```

```
hive> create table docs(line string);
OK
Time taken: 0.043 seconds
hive> load data inpath '/usr/data/word.txt' overwrite into table docs;
Loading data to table hive.docs
OK
Time taken: 0.164 seconds
hive> select * from docs;
OK
I love China
China is a great and beautiful country
I am Chinese
I am proud of being Chinese
Time taken: 0.267 seconds, Fetched: 4 row(s)
```

图7-60 创建数据表docs并装载数据

（2）分词

对文本内容进行分词，得到字符串数组，如图7-61所示，使用语句：

```
hive>select split(line,' ') from docs;
```

```
hive> select split(line,' ') from docs;
OK
["I","love","China"]
["China","is","a","great","and","beautiful","country"]
["I","am","Chinese"]
["I","am","proud","of","being","Chinese"]
Time taken: 0.133 seconds, Fetched: 4 row(s)
```

图7-61 分词

为方便对每个单词进行数据库操作，将得到的数组以列的形式组织，如图7-62所示。

```
hive> select explode(split(line,' ')) from docs;
```

> **注意**
>
> 任务前，需要提前创建数据库，例如，创建名为 hive 的数据库：
> hive> create database hive;

> **笔记**
>
> split(string str, string pat)：用于切分数据，按照 pat 将字符串 str 进行分隔，返回分隔后的字符串数组。

> **笔记**
>
> explode（array/map 类型）：将一个数组以列的形式组织。

```
hive> select explode(split(line,' ')) from docs;
OK
I
love
China
China
is
a
great
and
beautiful
country
I
am
Chinese
I
am
proud
of
being
Chinese
Time taken: 0.132 seconds, Fetched: 19 row(s)
```

图7-62　将字符串数组转换为列

（3）按单词分组并统计各分组数量

对单词进行分组，并统计每个分组的记录数，即单词出现次数，如图7-63所示。

```
>select word, count(1) as count from
>(select explode(split(line,' '))as word from docs) w
>group by word;
```

📝 笔记

count(1)：用于统计数据表中记录数。

📝 说明

(select explode(split(line,' '))as word from docs) w：生成临时表w，w表有一个字段word，存放经过分词处理的单词。

select word, count(1) as count from w group by word：对w表内的word字段进行分组，统计每个分组的记录数，生成结果，结果第一列为word，第二列为记录数。

```
hive> select word,count(1) as count from (select explode(split(line,'
')) as word from docs) w group by word;
Query ID = hadoop_20220823005416_a420edaa-20fa-483c-ba81-77f3578136bb
Total jobs = 1
Launching Job 1 out of 1
Number of reduce tasks not specified. Estimated from input data size:
1
In order to change the average load for a reducer (in bytes):
  set hive.exec.reducers.bytes.per.reducer=<number>
In order to limit the maximum number of reducers:
  set hive.exec.reducers.max=<number>
In order to set a constant number of reducers:
  set mapreduce.job.reduces=<number>
Job running in-process (local Hadoop)
2022-08-23 00:54:17,521 Stage-1 map = 100%,  reduce = 100%
Ended Job = job_local2010377344_0007
MapReduce Jobs Launched:
Stage-Stage-1:  HDFS Read: 3180 HDFS Write: 1084 SUCCESS
Total MapReduce CPU Time Spent: 0 msec
OK
China    2
Chinese  2
I        3
a        1
am       2
and      1
beautiful        1
being    1
country  1
great    1
is       1
love     1
of       1
proud    1
Time taken: 1.453 seconds, Fetched: 14 row(s)
```

图7-63　对单词分组并统计记录数

（4）排序

将结果以升序排序。

```
>select word, count(1) as count from
>(select explode(split(line,' '))as word from docs) w
>group by word
>order by word;
```

创建数据表word_count，用于保存词频统计的输出结果，如图7-64所示。

```
hive>create table word_count as
>select word, count(1) as count from
>(select explode(split(line,' '))as word from docs) w
>group by word
>order by word;
```

```
hive> create table word_count as
    > select word,count(1) as count from
    > (select explode(split(line,' '))as word from docs) w
    > group by word
    > order by word;
Query ID = hadoop_20220823011414_34ad404d-8fe2-43ab-915b-19189ddc05a6
Total jobs = 2
Launching Job 1 out of 2
Number of reduce tasks not specified. Estimated from input data size: 1
In order to change the average load for a reducer (in bytes):
  set hive.exec.reducers.bytes.per.reducer=<number>
In order to limit the maximum number of reducers:
  set hive.exec.reducers.max=<number>
In order to set a constant number of reducers:
  set mapreduce.job.reduces=<number>
Job running in-process (local Hadoop)
2022-08-23 01:14:16,143 Stage-1 map = 100%,  reduce = 100%
Ended Job = job_local1934076492_0014
Launching Job 2 out of 2
Number of reduce tasks determined at compile time: 1
In order to change the average load for a reducer (in bytes):
  set hive.exec.reducers.bytes.per.reducer=<number>
In order to limit the maximum number of reducers:
  set hive.exec.reducers.max=<number>
In order to set a constant number of reducers:
  set mapreduce.job.reduces=<number>
Job running in-process (local Hadoop)
2022-08-23 01:14:17,435 Stage-2 map = 100%,  reduce = 100%
Ended Job = job_local604040166_0015
Moving data to directory hdfs://localhost:9000/user/hive/warehouse/hive.db/word_count
MapReduce Jobs Launched:
Stage-Stage-1:  HDFS Read: 4296 HDFS Write: 1084 SUCCESS
Stage-Stage-2:  HDFS Read: 4296 HDFS Write: 1255 SUCCESS
Total MapReduce CPU Time Spent: 0 msec
OK
Time taken: 2.829 seconds
```

图7-64　词频统计并输出至表word_count

（5）查看结果

执行完成后，用select语句查看运行结果，如图7-65所示。

```
hive>select * from word_count;
```

```
hive> select * from word_count;
OK
China      2
Chinese    2
I          3
a          1
am         2
and        1
beautiful  1
being      1
country    1
great      1
is         1
love       1
of         1
proud      1
Time taken: 0.053 seconds, Fetched: 14 row(s)
```

图7-65　查看运行结果

> 说明
>
> 相当于使用一句指令，就完成了词频统计功能，非常方便。

拓展学习

1. Hive分区

庞大的数据集可能需要耗费大量的时间去处理。在许多场景下，可以通过分区或切片的方法减少每一次扫描总数据量，这种做法可以显著地改善性能。

扫码看视频

数据会依照单个或多个列进行分区，通常按照时间、地域或者是商业维度进行分区。例如video表，分区的依据可以是电影的种类和评级，另外，按照拍摄时间划分可能会得到更一致的结果。为了达到性能表现的一致性，对不同列的划分应该让数据尽可能均匀分布。最好的情况下，分区的划分条件能够对应where语句的部分查询条件。

Hive使用HDFS的子目录功能实现分区。每一个子目录包含了分区对应的列名和每一列的值。但是由于HDFS并不支持大量的子目录，这也给分区的使用带来了限制。所以有必要对表中的分区数量进行预估，从而避免因为分区数量过大而带来的一系列问题。

Hive查询通常使用分区的列作为查询条件。这样可以指定MapReduce任务在HDFS中指定的子目录下完成扫描的工作。HDFS的文件目录结构可以像索引一样进行高效利用。

2. Hive的架构

Hive的架构如图7-66所示，主要分为三个模块：

1）用户接口模块（UI）：包括CLI、HWI、JDBC、ODBC、Thrift Server等。CLI是Hive自带的一个命令行界面；HWI是Hive的一个简单网页界面；JDBC、ODBC以及Thrift Server可以向用户提供进行编程访问的接口。

2）驱动模块（Driver）：包括编译器、优化器、执行器等。所有命令和查询都会进入驱动模块，通过该模块对输入进行解析编译，对需求的计算进行优化，然后按照指定的步骤执行。

3）元数据存储模块（Metastore）：是一个独立的关系型数据库。通常是与MySQL数据库连接后创建的一个MySQL实例，也可以是Hive自带的Derby数据库实例。元数据存储模块中主要保存表模式和其他系统元数据，例如表的名称、表的列及其属性、表的分区及其属性、表的属性、表中数据所在位置信息等。

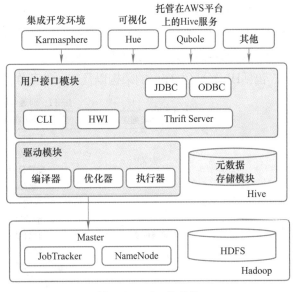

图7-66 Hive的架构

3. Hive的工作原理

如图7-67所示，Hive的工作原理为：

1）用户通过用户接口（UI）提交查询等任务给驱动程序（Driver）。

2）编译器（Compiler）获得该用户的任务Plan。

3）编译器（Compiler）根据用户任务去Metastore中获取需要的Hive的元数据信息。

4）编译器（Compiler）得到元数据信息，对任务进行编译，先将HiveQL转换为抽象语法树，然后将抽象语法树转换成查询块，将查询块转化为逻辑的查询计划，重写逻辑查询计划，将逻辑计划转化为物理的计划（MapReduce），最后选择最佳的策略。

5）将最终的计划提交给Driver。

6）Driver将计划（Plan）转交给执行引擎（Execution Engine）去执行。执行引擎获取元数据信息，提交给Hadoop中的JobTracker执行该任务，任务会直接读取HDFS中文件进行相应的操作。

7）用户请求获取执行的结果。

8）取得并返回执行结果。

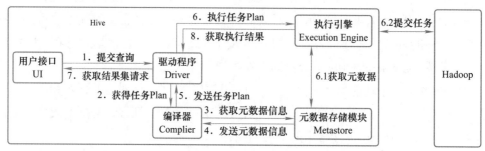

图7-67　Hive的工作原理

项目小结

本项目围绕"部署与使用Hive"典型工作任务，以大数据运维工程师新人视角，讲解了Hive、Hive与Hadoop、Hive部署模式、Hive与传统数据库、Hive的表类型及数据类型等基本概念，并由浅入深逐步展开四个任务，层层递进地学习如何部署与使用Hive，部署相关任务的横向对比如图7-68所示，引导大数据运维工程师新人逐步掌握部署与使用Hive的相关能力。在完成任务后，开展扩展知识学习，包括Hive分区、Hive的架构以及Hive的工作原理。

	任务1：部署本地模式Hive	任务2：部署远程模式Hive
环境配置	关闭防火墙	关闭防火墙
安装Hive	下载安装文件 解压安装文件 配置目录权限 配置PATH	下载安装文件 解压安装文件 配置目录权限 配置PATH
安装与部署MySQL	安装MySQL 配置MySQL 初始化MySQL 启动MySQL 配置hive库与hive用户	安装MySQL 配置MySQL 初始化MySQL 启动MySQL 配置hive库与hive用户
配置Hive	配置hive-env.sh文件 配置hive-default.xml文件 配置hive-site.xml文件 安装MySQL的JDBC驱动包	配置hive-env.sh文件 配置hive-default.xml文件 配置hive-site.xml文件 安装MySQL的JDBC驱动包 同步hive到slave1 配置目录权限 配置PATH 配置hive-site.xml文件
启动Hive	初始化元数据 启动Hadoop 启动Hive	初始化元数据 启动Hadoop 启动Metastore Server 启动Hive

图7-68　Hive部署任务横向对比总结

实战强化

1. 下载任务数据stocks.scv文件，进行如下实操练习。

1）在Hive中创建内部表stocks，字段分隔符为英文逗号，表结构见表7-6。

表7-6 表结构

col_name	data_type
exchange	string
symbol	string
ymd	string
price_open	float
price_high	float
price_low	float
price_close	float
volume	int
price_adj_close	float

2）将stocks.scv文件导入stocks表中，并查看数据。

3）从stocks表中查询苹果公司（symbol=AAPL）年平均调整后收盘价（price_adj_close）大于50美元的年份及年平均调整后收盘价。

2. 运维人员曾在Linux中卸载过MySQL，但因公司业务需求变化，需要在Linux中再次安装MySQL，在进行初始化密码的时候，运行结果如图7-69所示。

```
[root@master mysql]# mysqld --initialize-insecure --basedir=/usr
/local/mysql --datadir=/usr/local/mysql/data --user=root
2022-10-27T14:00:39.465100Z 0 [Warning] TIMESTAMP with implicit
DEFAULT value is deprecated. Please use --explicit_defaults_for_
timestamp server option (see documentation for more details).
2022-10-27T14:00:39.490097Z 0 [ERROR] --initialize specified but
 the data directory has files in it. Aborting.
2022-10-27T14:00:39.490974Z 0 [ERROR] Aborting
```

图7-69 安装MySQL出错

试分析运维人员出错的原因，并给出解决方案。

3. 运维人员在进行Hive环境部署时需要启动MySQL服务，在启动MySQL服务时，运行结果如图7-70所示。

```
[root@master mysql]# service mysql start
Starting MySQL...... ERROR! The server quit without updating PID
 file (/usr/local/mysql/data/master.pid).
```

图7-70 启动MySQL出错

试分析运维人员出错的原因，并给出解决方案。

4. 运维人员在服务器上部署了本地模式Hive，使用指令hive启动Hive，系统反馈如图7-71所示错误信息。

```
Exception in thread "main" java.lang.RuntimeException: java.net.ConnectException: Call From localhost/127.0.0.1 to localhost:9000 failed on connection exception: java.net.ConnectException: 拒绝连接; For more details see:   http://wiki.apache.org/hadoop/ConnectionRefused
        at org.apache.hadoop.hive.ql.session.SessionState.start(SessionState.java:651)
        at org.apache.hadoop.hive.ql.session.SessionState.beginStart(SessionState.java:591)
        at org.apache.hadoop.hive.cli.CliDriver.run(CliDriver.java:747)
        at org.apache.hadoop.hive.cli.CliDriver.main(CliDriver.java:683)
        at sun.reflect.NativeMethodAccessorImpl.invoke0(Native Method)
        at sun.reflect.NativeMethodAccessorImpl.invoke(NativeMethodAccessorImpl.java:62)
        at sun.reflect.DelegatingMethodAccessorImpl.invoke(DelegatingMethodAccessorImpl.java:43)
        at java.lang.reflect.Method.invoke(Method.java:498)
        at org.apache.hadoop.util.RunJar.run(RunJar.java:323)
        at org.apache.hadoop.util.RunJar.main(RunJar.java:236)
```

图7-71　启动Hive出错

试分析运维人员出错的原因，并给出解决方案。

项目 8　部署与使用 Spark

项目概述

随着数据规模的不断扩大和对数据处理要求的不断提高，A公司之前部署的基于Hadoop的大数据平台项目对某些任务的处理无法达到相关要求。现公司委派大数据运维工程师王工前往客户的中心机房对平台进行Spark集群部署，并为客户演示Spark的使用。

本项目针对"部署与使用Spark"的典型工作任务，依次完成四个工作任务：部署单机模式Spark、部署Spark集群、使用Spark Shell编写代码、使用Scala编写Spark程序。

在开展任务前，需要掌握必要的理论知识：什么是Spark？什么是Spark的生态系统？Spark与Hadoop的区别是什么？Spark的优势是什么？三种部署模式分别指的是什么？Spark Shell是什么？

完成任务后，进一步了解与Spark架构相关的知识，加深对Spark架构的理解。

学习目标

1. 了解Spark的简介与生态。
2. 了解Spark与Hadoop的关系。
3. 了解Spark的部署方式。
4. 能按照流程部署Spark。
5. 能使用Spark Shell编写代码。
6. 能使用Scala编写Spark程序。
7. 了解Spark运行框架。

思维导图

项目思维导图如图8-1所示。

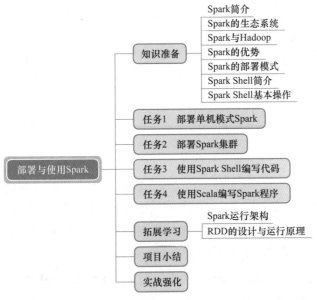

图8-1　项目思维导图

知识准备

1. Spark简介

扫码看视频

Spark最初由美国加州伯克利大学（UC Berkeley）的AMP实验室于2009年开发，是基于内存计算的大数据并行计算框架，可用于构建大型的、低延迟的数据分析应用程序。

Spark具有如下几个主要特点：

1）运行速度快：使用DAG执行引擎以支持循环数据流与内存计算。

2）容易使用：支持使用Scala、Java、Python和R语言进行编程，可以通过Spark Shell进行交互式编程。

3）通用性：Spark提供了完整而强大的技术栈，包括SQL查询、流式计算、机器学习和图算法组件。

4）运行模式多样：可运行于独立的集群模式中，可运行于Hadoop中，也可运行于Amazon EC2等云环境中，并且可以访问HDFS、Cassandra、HBase、Hive等多种数据源。

2. Spark的生态系统

Spark的设计遵循"一个软件栈满足不同应用场景"的理念，逐渐形成了一套完整的生态系统，既能够提供内存计算框架，也可以支持SQL即席查询、实时流式计算、机器学习和图计算等。Spark可以部署在资源管理器YARN之上，提供一站式的大数据解决方案。因此，Spark所提供的生态系统足以应对批处理、交互式查询和流数据处理三种场景。

Spark的生态系统主要包含了Spark Core、Spark SQL、Spark Streaming、MLLib和GraphX等组件，具体功能见表8-1。

表8-1 Spark生态系统中组件的具体功能

组 件	功 能
Spark Core	提供内存计算
Spark SQL	提供交互式查询分析
Spark Streaming	提供流计算功能
MLLib	提供机器学习算法库的组件
GraphX	提供图计算

3. Spark与Hadoop

Hadoop存在许多缺点：①表达能力有限；②磁盘IO开销大；③延迟高；④任务之间的衔接涉及IO开销；⑤在前一个任务执行完成之前，其他任务就无法开始，难以胜任复杂、多阶段的计算任务。

Spark在借鉴Hadoop MapReduce优点的同时，很好地解决了MapReduce所面临的问题。相比于Hadoop MapReduce，Spark主要具有如下优点：

1）Spark的计算模式也属于MapReduce，但不局限于Map和Reduce操作，还提供了多种数据集操作类型，编程模型比Hadoop MapReduce更灵活。

2）Spark提供了内存计算，可将中间结果放到内存中，对于迭代运算效率更高。

3）Spark基于DAG的任务调度执行机制，要优于Hadoop MapReduce的迭代执行机制。

4. Spark的优势

Spark具有以下优势：

1）更快的速度：内存计算下，Spark比Hadoop快100倍。

2）易用性：Spark提供了80多个高级运算符。

3）通用性：Spark提供了大量的库，包括SQL、DataFrames、MLlib、GraphX、Spark Streaming。开发者可以在同一个应用程序中无缝组合使用这些库。

5. Spark的部署模式

Spark支持四种不同类型的部署模式，包括：

1）Local：单机模式，常用于本地测试。

2）Standalone：类似于MapReduce1.0，slot为资源分配单位。

3）Spark on Mesos：和Spark有血缘关系，更好支持Mesos。

4）Spark on YARN。

6. Spark Shell简介

学习Spark程序开发，建议首先通过Spark Shell进行交互式编程，加深对Spark程序开发的理解。

Spark Shell提供了简单的方式来学习Spark API，并且可以以实时、交互的方式来分析数据。开发人员可以输入一条语句，Spark Shell会立即执行语句并返回结果，而不必等到整个程序运行完毕，因此可以及时查看中间结果并对程序进行修改，这样可以在很大程度上提升程序开发效率。

Spark Shell支持Scala和Python。由于Spark框架本身使用Scala语言开发的，因此，使用spark-shell命令默认进入Scala的交互式执行环境。如果要进入Python的交互式执行环境，则需要执行pyspark命令。

7. Spark Shell基本操作

Spark对于分布式数据集最主要的抽象为弹性分布式集合（Resilient Distributed Dataset，RDD）。见表8-2、表8-3，Spark RDD支持两种类型的操作：

1）动作（action）：在数据集上进行运算，返回计算值。

2）转换（transformation）：基于现有的数据集创建一个新的数据集。

扫码看视频

表8-2　动作操作

Action API	说　　明
count()	返回数据集中的元素个数
collect()	以数组的形式返回数据集中的所有元素
first()	返回数据集中的第一个元素
take(n)	以数组的形式返回数据集中的前n个元素
reduce(func)	通过函数func（输入两个参数并返回一个值）聚合数据集中的元素
foreach(func)	将数据集中的每个元素传递到函数func中运行

表8-3 转换操作

Transformation API	说 明
filter(func)	筛选出满足函数func的元素,并返回一个新的数据集
map(func)	将每个元素传递到函数func中,并将结果返回为一个新的数据集
flatMap(func)	与map()相似,但每个输入元素都可以映射到0或多个输出结果
groupByKey()	应用于(K, V)键值对的数据集时,返回一个新的(K, Iterable<V>)形式的数据集
reduceByKey(func)	应用于(K, V)键值对的数据集时,返回一个新的(K, V)形式的数据集,其中的每个值是将每个key传递到函数func中进行聚合

任务1 部署单机模式Spark

扫码看视频

任务描述

王工在前往客户现场前,为了确保顺利实现Spark的集群部署,先在自己公司的单机环境下进行单机模式部署,并测试Spark的基础功能。

任务分析

1. 任务目标

1)了解Spark概念、生态系统及运行架构。
2)掌握Spark单机部署。
3)能够运行Spark应用程序。

2. 任务环境

操作系统:CentOS Stream 9(预装分布式Hadoop)
软件版本:JDK 1.8、Hadoop 3.3.4、Spark 3.3.0

3. 任务导图

任务导图如图8-2所示。

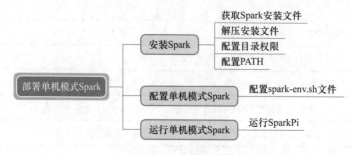

图8-2 任务导图

任务实施

1. 安装Spark

（1）获取Spark安装文件

打开浏览器，访问Apache官方下载路径：https://archive.apache.org，找到Spark对应版本的下载路径，使用指令下载安装包，例如spark-3.3.0-bin-hadoop3.tgz，如图8-3所示。

```
# cd /usr/local
# wget https://archive.apache.org/dist/spark/spark-3.3.0/spark-3.3.0-bin-hadoop3.tgz
```

> 📝 **说明**
> 本任务需要在分布式hadoop环境基础上完成。

> 📝 **说明**
> 随着版本更新，当前版本号的下载地址可能发生变化，请读者自行到官网获取相应下载地址。

> 📝 **说明**
> 如果通过wget指令获取安装包太慢，可以使用本书提供的安装包。使用scp指令从Windows传输文件至Linux中。

> ⚠️ **注意**
> 下载过程需保持网络状态。

```
[root@localhost ~]# cd /usr/local
[root@localhost local]# wget https://archive.apache.org/dist/spark/spark-3.3.0/spark-3.3.0-bin-hadoop3.tgz
--2022-08-24 02:58:28--  https://archive.apache.org/dist/spark/spark-3.3.0/spark-3.3.0-bin-hadoop3.tgz
正在解析主机 archive.apache.org (archive.apache.org)... 198.18.6.45
正在连接 archive.apache.org (archive.apache.org)|198.18.6.45|:443... 已连接。
已发出 HTTP 请求，正在等待回应... 200 OK
长度：299321244 (285M) [application/x-gzip]
正在保存至: "spark-3.3.0-bin-hadoop3.tgz"

spark-3.3.0-bin-hadoop 100%[=======================>] 285.45M  17.8MB/s  用时 19s
```

图8-3 下载Spark

（2）解压安装文件

```
# tar -zxf spark-3.3.0-bin-hadoop3.tgz
```

更改目录名称为spark。

```
# mv spark-3.3.0-bin-hadoop3 spark
```

（3）配置目录权限

将目录的所有者改为hadoop用户，如图8-4所示。

```
# chown -R hadoop: hadoop spark
```

```
[root@localhost local]# tar -zxf spark-3.3.0-bin-hadoop3.tgz
[root@localhost local]# mv spark-3.3.0-bin-hadoop3 spark
[root@localhost local]# chown -R hadoop:hadoop spark
```

图8-4 解压、重命名、修改权限

（4）配置PATH

使用vim编辑器编辑/etc/profile。

```
# vim /etc/profile
```

在打开的profile文件中，添加Spark相关路径。

在末行增加如下内容，如图8-5所示。

```
export SPARK_HOME=/usr/local/spark
export PATH=$PATH:$SPARK_HOME/bin
```

```
export JAVA_HOME=/usr/lib/jvm/java-1.8.0-openjdk-1.8.0.345.b01-2.el9.x86_64/jre
export PATH=$PATH:$JAVA_HOME

export HADOOP_HOME=/usr/local/hadoop
export PATH=$PATH:$HADOOP_HOME/bin:$HADOOP_HOME/sbin

export SPARK_HOME=/usr/local/spark
export PATH=$PATH:$SPARK_HOME/bin
```

图8-5 配置/etc/profile

大数据平台部署与运维

> ⚠ **注意**
> 更新 /etc/profile 后务必记得使用 source 指令使更新后的内容生效。

> 📝 **说明**
> 通过前面的步骤,已经设置 hadoop 用户具有 spark 文件夹的权限,因此,后面的步骤由 hadoop 用户执行。

> 📝 **笔记**
> 有了 spark-env.sh 配置信息以后,Spark 就可以把数据存储到 Hadoop 分布式文件系统 HDFS 中,也可以从 HDFS 中读取数据。如果没有配置 spark-env.sh,Spark 就只能读写本地数据,无法读写 HDFS 中的数据。
> /usr/local/hadoop/bin/hadoop classpath 其实是一条脚本,执行结果返回的是一系列路径,路径下存储的是 Hadoop 所依赖的 jar 包。将其配置给 Spark,使 Spark 得以架构在 Hadoop 之上,调用 HDFS。

配置完成后,按<ESC>键,输入:wq,保存并退出。

使用 source 指令,使新配置生效。

```
# source /etc/profile
```

2. 配置单机模式 Spark

切换到 hadoop 用户。

```
# su - hadoop
```

配置 spark-env.sh 文件

复制 Spark 安装文件自带的配置文件模板 spark-env.sh.template,作为 Spark 配置文件 spark-env.sh,如图 8-6 所示。

```
$ cd /usr/local/spark
$ cp ./conf/spark-env.sh.template ./conf/spark-env.sh
```

```
[root@localhost spark]# su - hadoop
[hadoop@localhost ~]$ cd /usr/local/spark/
[hadoop@localhost spark]$ cp ./conf/spark-env.sh.template ./conf/spark-env.sh
```

图 8-6 复制 Spark 配置文件模板

编辑 /usr/local/spark/conf/spark-env.sh 文件,在该文件的第一行添加以下配置信息,如图 8-7 所示。

```
export
SPARK_DIST_CLASSPATH=$(/usr/local/hadoop/bin/hadoop classpath)
```

```
export SPARK_DIST_CLASSPATH=$(/usr/local/hadoop/bin/hadoop classpath)
#!/usr/bin/env bash
```

图 8-7 修改配置文件 spark-env.sh

3. 运行单机模式 Spark

运行 SparkPi

如图 8-8 所示,运行 Spark 自带的实例,可以验证 Spark 是否安装成功,命令如下:

```
$ run-example SparkPi
```

```
22/09/17 20:57:52 INFO TaskSchedulerImpl: Removed TaskSet 0.0, whose tasks have all completed, from pool
22/09/17 20:57:52 INFO DAGScheduler: Job 0 is finished. Cancelling potential spe
culative or zombie tasks for this job
22/09/17 20:57:52 INFO TaskSchedulerImpl: Killing all running tasks in stage 0:
Stage finished
22/09/17 20:57:52 INFO DAGScheduler: Job 0 finished: reduce at SparkPi.scala:38,
 took 2.231412 s
Pi is roughly 3.142915714578573
22/09/17 20:57:52 INFO SparkUI: Stopped Spark web UI at http://master:4040
22/09/17 20:57:52 INFO MapOutputTrackerMasterEndpoint: MapOutputTrackerMasterEnd
point stopped!
22/09/17 20:57:52 INFO MemoryStore: MemoryStore cleared
22/09/17 20:57:52 INFO BlockManager: BlockManager stopped
22/09/17 20:57:52 INFO BlockManagerMaster: BlockManagerMaster stopped
22/09/17 20:57:52 INFO OutputCommitCoordinator$OutputCommitCoordinatorEndpoint:
OutputCommitCoordinator stopped!
22/09/17 20:57:52 INFO SparkContext: Successfully stopped SparkContext
22/09/17 20:57:52 INFO ShutdownHookManager: Shutdown hook called
22/09/17 20:57:52 INFO ShutdownHookManager: Deleting directory /tmp/spark-786d24
6e-0634-4801-bc97-64eea2c62f83
22/09/17 20:57:52 INFO ShutdownHookManager: Deleting directory /tmp/spark-49e14b
57-8a7e-4f53-a1a1-452017f7d56d
```

图 8-8 运行 SparkPi

执行时会输出很多屏幕信息，不容易找到最终的输出结果，为了从大量的输出信息中快速找到想要的执行结果，可以通过grep命令进行过滤，如图8-9所示。

```
$ run-example SparkPi 2>&1 | grep "Pi is roughly"
```

```
[hadoop@localhost spark]$ run-example SparkPi 2>&1|grep "Pi is roughly"
Pi is roughly 3.142915714578573
```

图8-9　grep命令过滤后的SparkPi运行结果

> **笔记**
> 2>&1：1表示标准输出，2表示标准错误输出，2>&1表示将标准错误输出重定向到标准输出中。

任务2　部署Spark集群

扫码看视频

任务描述

王工来到客户现场，客户的集群环境中有全分布模式Hadoop了，接下来，为客户部署完全分布式Spark集群。

任务分析

1. 任务目标

1）掌握Spark完全分布式部署。
2）能够运行Spark应用程序。

2. 任务环境

操作系统：CentOS Stream 9（预装全分布模式Hadoop）
软件版本：JDK 1.8、Hadoop 3.3.4、Spark 3.3.0

3. 任务导图

任务导图如图8-10所示。

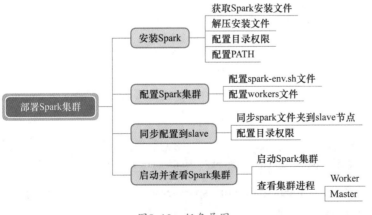

图8-10　任务导图

任务实施

1. 安装Spark

（1）获取Spark安装文件

打开浏览器，访问Apache官方下载路径：https://archive.apache.org，找到Spark对应版本的下载路径，使用指令下载安装包，例如spark-3.3.0-bin-hadoop3.tgz。

```
# cd /usr/local
#wget https://archive.apache.org/dist/spark/spark-3.3.0/spark-3.3.0-bin-hadoop3.tgz
```

（2）解压安装文件

```
# tar -zxf spark-3.3.0-bin-hadoop3.tgz
```

更改目录名称为spark。

```
# mv spark-3.3.0-bin-hadoop3 spark
```

（3）配置目录权限

将目录的所有者改为hadoop用户。

```
# chown -R hadoop:hadoop spark
```

（4）配置PATH

使用vim编辑器编辑/etc/profile。

```
# vim /etc/profile
```

在打开的profile文件中，添加Spark相关路径。

在末行，增加如下内容：

```
export SPARK_HOME=/usr/local/spark
export PATH=$PATH:$SPARK_HOME/bin
```

使用source指令，使新配置生效。

```
# source /etc/profile
```

2. 配置Spark集群

切换到hadoop用户。

```
# su - hadoop
```

（1）配置spark-env.sh文件

复制Spark安装文件自带的配置文件模板spark-env.sh.template，作为Spark配置文件spark-env.sh。

```
$ cd /usr/local/spark
$ cp ./conf/spark-env.sh.template ./conf/spark-env.sh
```

编辑/usr/local/spark/conf/spark-env.sh文件，在该文件的第一行添加以下配置信息：

说明

本任务需要在完全分布式Hadoop环境上完成。

说明

安装Spark过程与任务1相同，详细内容参照任务1，此处仅罗列必要步骤。

说明

随着版本更新，当前版本号的下载地址可能发生变化，请读者自行到官网获取相应下载地址。

说明

如果通过wget指令获取安装包太慢，可以使用本书提供的安装包。使用scp指令从Windows传输文件至Linux中。

注意

下载过程需保持网络状态。

注意

更新/etc/profile后务必使用source指令使更新后的内容生效。

说明

配置Spark集群过程（1）与任务1相同，详细内容参照任务1，此处仅罗列必要步骤。

笔记

有了spark-env.sh配置信息以后，Spark就可以把数据存储到Hadoop分布式文件系统HDFS中，也可以从HDFS中读取数据。如果没有配置spark-env.sh，Spark就只能读写本地数据，无法读写HDFS中的数据。

```
export SPARK_DIST_CLASSPATH=$(/usr/local/hadoop/bin/hadoop classpath)
```

(2)配置workers文件

在/usr/local/spark/conf/路径下,新建workers文件。

```
$ cd /usr/local/spark/conf
$ vi workers
```

配置以下内容,如图8-11所示。

```
master
slave1
slave2
```

> **笔记**
> /usr/local/hadoop/bin/hadoop classpath 其实是一条脚本,执行结果返回的是一系列路径,路径下存储的是 Hadoop 所依赖的 jar 包。将其配置给 Spark,使 Spark 得以架构在 Hadoop 之上,调用 HDFS。

```
[hadoop@master conf]$ cd /usr/local/spark/conf/
[hadoop@master conf]$ vi workers
[hadoop@master conf]$ cat workers
master
slave1
slave2
```

图8-11 配置workers

> **说明**
> workers 文件也可以从 /usr/local/spark/conf 路径下的 workers.template 文件复制得到并修改配置内容,效果相同。

3. 同步配置到slave

(1)同步spark文件夹到slave节点

将master上的spark文件夹同步到slave1、slave2节点,图8-12所示为复制至slave1节点,slave2节点类似。

```
$ scp -r /usr/local/spark/ root@slave1:/usr/local/
$ scp -r /usr/local/spark/ root@slave2:/usr/local/
```

```
[hadoop@master conf]$ scp -r /usr/local/spark root@slave1:/usr/local
root@slave1's password:
NOTICE                              100%   56KB  20.8MB/s   00:00
python_executable_check.py          100%  1513   2.7MB/s   00:00
autoscale.py                        100%  1592   3.4MB/s   00:00
worker_memory_check.py              100%  1585   3.5MB/s   00:00
```

图8-12 在master节点复制spark文件夹至slave1

(2)配置目录权限

在slave1、slave2节点修改spark文件夹的所属者为hadoop用户。

```
# chown -R hadoop:hadoop /usr/local/spark
```

4. 启动并查看Spark集群

(1)启动Spark集群

在spark目录下执行命令,如图8-13所示。

```
$ cd /usr/local/spark
$ sbin/start-all.sh
```

```
[hadoop@master spark]$ sbin/start-all.sh
starting org.apache.spark.deploy.master.Master, logging to /usr/local/spark/logs
/spark-hadoop-org.apache.spark.deploy.master.Master-1-master.out
slave2: starting org.apache.spark.deploy.worker.Worker, logging to /usr/local/sp
ark/logs/spark-hadoop-org.apache.spark.deploy.worker.Worker-1-slave2.out
slave1: starting org.apache.spark.deploy.worker.Worker, logging to /usr/local/sp
ark/logs/spark-hadoop-org.apache.spark.deploy.worker.Worker-1-slave1.out
master: starting org.apache.spark.deploy.worker.Worker, logging to /usr/local/sp
ark/logs/spark-hadoop-org.apache.spark.deploy.worker.Worker-1-master.out
```

图8-13 启动Spark集群

（2）查看集群进程

启动成功后，使用jps查看进程，如图8-14所示。

```
[hadoop@master spark]$ jps
15777 Worker
15627 Master
15838 Jps
        a)

[hadoop@slave1 ~]$ jps
7869 Worker
8015 Jps
[hadoop@slave1 ~]$
        b)

[hadoop@slave2 ~]$ jps
7361 Jps
7178 Worker
[hadoop@slave2 ~]$
        c)
```

图8-14　使用jps查看集群进程

a）master节点　b）slave1节点　c）slave2节点

任务3　使用Spark Shell编写代码

扫码看视频

任务描述

王工为客户部署完全分布式Spark集群后，需要测试Spark集群的可用性。因此本任务中，王工使用Spark Shell编写代码，实现HDFS文件处理与词频统计功能。

任务分析

1. 任务目标

1）掌握Spark Shell的启动方式。

2）熟悉Spark Shell的基本操作。

2. 任务环境

操作系统：CentOS Stream 9（预装分布式Hadoop+Spark）

软件版本：Java 1.8.0、Hadoop 3.3.4、Spark 3.3.0

3. 任务导图

任务导图如图8-15所示。

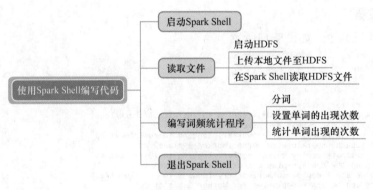

图8-15　任务导图

任务实施

1. 启动Spark Shell

启动Spark Shell环境。

```
$ spark-shell
```

启动Spark Shell后就会进入"scala>"命令提示符状态，如图8-16所示。

```
Spark context available as 'sc' (master = local[*], app id = local-1661
).
Spark session available as 'spark'.
Welcome to
      ____              __
     / __/__  ___ _____/ /__
    _\ \/ _ \/ _ `/ __/  '_/
   /___/ .__/\_,_/_/ /_/\_\   version 3.3.0
      /_/

Using Scala version 2.12.15 (OpenJDK 64-Bit Server VM, Java 1.8.0_332)
Type in expressions to have them evaluated.
Type :help for more information.
```

图8-16　Spark Shell环境

> **说明**
>
> 启动 Spark Shell 时，不需要启动 Hadoop，即运行 Spark 是不依赖 Hadoop 的。

> **注意**
>
> 使用 hadoop 用户启动 Spark Shell。

现在就可以输入Scala代码进行调试了。如图8-17所示，在Scala命令提示符"scala>"后面输入一个表达式"8*2+5"，然后按<Enter>键，就会立即得到结果。

```
scala> 8*2+5
res0: Int = 21
```

图8-17　Scala简单调试

2. 读取文件

（1）启动HDFS

启动HFDS。

```
$ start-dfs.sh
```

（2）上传本地文件至HDFS

把本地文件"/usr/local/spark/README.md"上传到HDFS的"/user/hadoop"目录下，并使用cat命令输出HDFS中的README.md内容，如图8-18所示，说明文件上传成功。

```
# hdfs dfs -put /usr/local/spark/README.md .
# hdfs dfs -cat README.md
```

> **注意**
>
> 在上传文件前，要确保 HDFS 中有"/user/hadoop"目录，若没有，可使用命令进行创建：hdfs dfs-mkdir-p/user/hadoop。

```
[hadoop@master ~]$ hdfs dfs -mkdir -p /user/hadoop
[hadoop@master ~]$ hdfs dfs -put /usr/local/spark/README.md .
[hadoop@master ~]$ hdfs dfs -ls .
Found 1 items
-rw-r--r--   3 hadoop supergroup       4461 2022-11-02 08:06 README.md
[hadoop@master ~]$ hdfs dfs -cat README.md
# Apache Spark

Spark is a unified analytics engine for large-scale data processing. It provides
high-level APIs in Scala, Java, Python, and R, and an optimized engine that
```

图8-18　往HDFS上传文件

ⓘ 扩展

读取本地文件：读取 Linux 本地文件系统中的文件并显示第一行的内容，如图 8-20 所示。

```
scala> val textFile = sc.textFile
("file:///usr/local/spark/README.
md")
scala> textFile.first()
```

```
scala> var textFile=sc.textFile("file:///usr/local/spark/
README.md")
textFile: org.apache.spark.rdd.RDD[String] = file:///usr/
local/spark/README.md MapPartitionsRDD[3] at textFile at
<console>:23

scala> textFile.first()
res1: String = # Apache Spark
```

图 8-20　读取本地文件结果

📝 笔记

1）textFile 包含多行文本内容。

2）textFile.flatMap(line => line.split(" ")) 会遍历 textFile 中的每行文本内容，当遍历到其中一行文本内容时，会把文本内容赋值给变量 line，并执行 Lamda 表达式。

line => line.split(" ") 是一个 Lamda 表达式，左边表示输入参数，右边表示函数里面执行的处理逻辑，这里执行 line.split(" ")，也就是针对 line 中的一行文本内容，采用空格作为分隔符进行单词切分，从一行文本切分得到很多个单词构成的单词集合。这样，对于 textFile 中的每行文本，都会使用 Lamda 表达式得到一个单词集合，最终对多行文本使用 Lamda 表达式，就得到多个单词集合。

3）textFile.flatMap() 操作把多个单词集合合并得到一个大的单词集合。

4）collect() 是为了在界面中以数组展示上述操作结果。

（3）在 Spark Shell 读取 HDFS 文件

现在切换到之前已经打开的 Spark Shell 窗口，编写语句从 HDFS 中加载 README.md 文件，并显示第一行文本内容，如图 8-19 所示。

```
scala> val textFile = sc.textFile("hdfs://master:9000/user/hadoop/README.md")
scala> textFile.first()
```

```
scala> val textFile = sc.textFile("hdfs://master:9000/user/hadoop/READ
ME.md")
textFile: org.apache.spark.rdd.RDD[String] = hdfs://master:9000/user/h
adoop/README.md MapPartitionsRDD[3] at textFile at <console>:23

scala> textFile.first()
res1: String = # Apache Spark
```

图 8-19　读取 HDFS 文件结果

执行上面语句后，就可以看到 HDFS 文件系统中（不是本地文件系统）的 README.md 的第一行内容了，即"# Apache Spark"。

3. 编写词频统计程序

（1）分词

对文本内容分词，如图 8-21 所示。

```
scala> textFile.flatMap(line=>line.split(" ")).collect()
```

```
scala> textFile.flatMap(line=>line.split(" ")).collect()
res5: Array[String] = Array(#, Apache, Spark, "", Spark, is, a, uni
fied, analytics, engine, for, large-scale, data, processing., It, p
rovides, high-level, APIs, in, Scala,, Java,, Python,, and, R,, and
, an, optimized, engine, that, supports, general, computation, grap
hs, for, data, analysis., It, also, supports, a, rich, set, of, hig
her-level, tools, including, Spark, SQL, for, SQL, and, DataFrames,
, pandas, API, on, Spark, for, pandas, workloads,, MLlib, for, mach
ine, learning,, GraphX, for, graph, processing,, and, Structured, S
treaming, for, stream, processing., "", <https://spark.apache.org/>
, "", [![GitHub, Action, Build](https://github.com/apache/spark/act
ions/workflows/build_and_test.yml/badge.svg?branch=master&event=pus
h)](https://github.com/apache/spar...
```

图 8-21　分词

（2）设置单词的出现次数

为每个单词设置出现次数为 1，如图 8-22 所示。

```
val wordCount = textFile.flatMap(line => line.split(" ")).map(word => (word, 1)).collect()
```

```
scala> textFile.flatMap(line=>line.split(" ")).map(word=>(word,1)).
collect()
res6: Array[(String, Int)] = Array((#,1), (Apache,1), (Spark,1), ("
",1), (Spark,1), (is,1), (a,1), (unified,1), (analytics,1), (engine
,1), (for,1), (large-scale,1), (data,1), (processing.,1), (It,1), (
provides,1), (high-level,1), (APIs,1), (in,1), (Scala,,1), (Java,,1
), (Python,,1), (and,1), (R,,1), (and,1), (an,1), (optimized,1), (e
ngine,1), (that,1), (supports,1), (general,1), (computation,1), (gr
aphs,1), (for,1), (data,1), (analysis.,1), (It,1), (also,1), (suppo
rts,1), (a,1), (rich,1), (set,1), (of,1), (higher-level,1), (tools,
,1), (including,1), (Spark,1), (SQL,1), (for,1), (SQL,1), (and,1), (
DataFrames,,1), (pandas,1), (API,1), (on,1), (Spark,1), (for,1), (p
andas,1), (workloads,,1), (MLlib,1), (for,1), (machine,1), (learnin
g,,1), (GraphX,1), (for,1), (graph...
```

图8-22　设置单词出现次数为1

笔记

针对大的单词集合，执行map()操作，这一操作会遍历这个集合中的每个单词，当遍历到其中一个单词时，就把当前这个单词赋值给变量word，并执行Lamda表达式。

"word => (word, 1)"：word作为函数的输入参数，然后执行函数处理逻辑，这里会执行(word, 1)，也就是针对输入的word，构建得到一个映射，这个映射的key是word，value是1。

（3）统计单词出现的次数

程序执行到这时，已经得到一个映射(Map)，这个映射中包含了很多(key, value)。最后针对这个映射，执行聚合操作，返回聚合后的(key, value)，如图8-23所示。

笔记

reduceByKey((a, b) => a + b)：把映射中的所有(key, value)按照key进行分组，然后使用给定的函数（即Lamda表达式：(a, b) => a + b）。对具有相同的key的多个value进行聚合操作，即对相同单词的value值相加，得到该单词出现的总次数。

```
scala> textFile.flatMap(line => line.split(" ")).map(word => (word, 1)).
reduceByKey((a, b) => a + b) .collect()
```

```
scala> textFile.flatMap(line=>line.split(" ")).map(word=>(word,1)).
reduceByKey((a,b)=>a+b).collect()
res7: Array[(String, Int)] = Array((package,1), (this,1), (integrat
ion,1), (Python,2), (cluster.,1), (its,1), ([run,1), (There,1), (ge
neral,2), (have,1), (pre-built,1), (Because,1), (YARN,,1), (locally
,2), (changed,1), (locally.,1), (several,1), (only,1), (Configurati
on,1), (This,2), (basic,1), (first,1), (learning,,1), (documentatio
n,3), (graph,1), (Hive,2), (info,1), (["Specifying,1), ("yarn",1),
([params]`.,1), ([project,1), (prefer,1), (SparkPi,2), (engine,2),
(version,1), (file,1), (documentation,,1), (Action,1), (MASTER,1),
(example,3), (are,1), (systems.,1), (params,1), (scala>,1), (DataFr
ames,,1), (provides,1), (refer,2), (configure,1), (Interactive,2),
(R,,1), (can,6), (build,3), (when,1), (easiest,1), (Maven](https://
maven.apache.org/).,1), (Apache,1)...
```

图8-23　词频统计程序实现

4. 退出Spark Shell

使用命令":quit"退出Spark Shell，如图8-24所示。

```
scala>:quit
```

笔记

也可以直接使用快捷键<Ctrl+D>退出Spark Shell。

```
scala> :quit
[hadoop@localhost ~]$
```

图8-24　退出Spark Shell环境

任务4　使用Scala编写Spark程序

扫码看视频

任务描述

王工为客户部署完全分布式Spark集群后，需要测试Spark集群的可用性。王工在Spark环境中编译Scala语言编写的Spark应用程序，并编译、打包、执行。

任务分析

1. 任务目标

1）熟练掌握Spark应用的提交方法。

2）能够使用Scala语言编写Spark独立应用程序。

3）能够使用Maven工具对Spark应用程序进行编译打包。

2. 任务环境

操作系统：CentOS Stream 9（预装分布式Hadoop+Spark）

软件版本：Java 1.8.0、Hadoop 3.3.4、Spark 3.3.0、Maven 3.8.6

3. 任务导图

任务导图如图8-25所示。

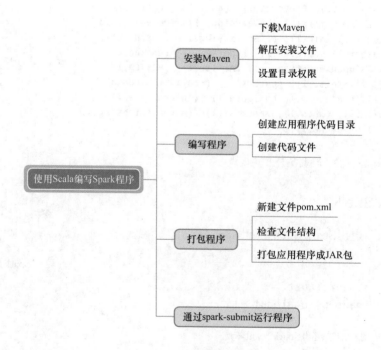

图8-25　任务导图

任务实施

1. 安装Maven

（1）下载Maven

打开浏览器，访问Maven官网（https://downloads.apache.org/maven/maven-3/3.8.6/binaries/apache-maven-3.8.6-bin.tar.gz），下载安装文件apache-maven-3.8.6-bin.tar.gz，如图8-26所示。

```
# cd /usr/local
# wget https://downloads.apache.org/maven/maven-3/3.8.6/binaries/apache-maven-3.8.6-bin.tar.gz
```

```
[root@localhost ~]# cd /usr/local
[root@localhost local]# wget https://downloads.apache.org/maven/maven-3/3.8.6/binaries/apache-maven-3.8.6-bin.tar.gz
--2022-08-28 07:46:45--  https://downloads.apache.org/maven/maven-3/3.8.6/binaries/apache-maven-3.8.6-bin.tar.gz
正在解析主机 downloads.apache.org (downloads.apache.org)... 198.18.3.116
正在连接 downloads.apache.org (downloads.apache.org)|198.18.3.116|:443... 已连接。
已发出 HTTP 请求，正在等待回应... 200 OK
长度： 8676320 (8.3M) [application/x-gzip]
正在保存至: "apache-maven-3.8.6-bin.tar.gz"

apache-maven-3.8 100%[===========>]   8.27M  2.55MB/s    用时 3.2s

2022-08-28 07:46:50 (2.55 MB/s) - 已保存 "apache-maven-3.8.6-bin.tar.gz" [8676320/8676320]
```

图8-26　下载Maven

（2）解压安装文件

解压安装包，并为文件夹重命名为maven。

```
# tar -zxf apache-maven-3.8.6-bin.tar.gz
# mv apache-maven-3.8.6 maven
```

（3）设置目录权限

将目录的所有者改为hadoop用户，如图8-27所示。

```
# chown -R hadoop:hadoop maven
```

```
[root@master local]# tar -zxf apache-maven-3.8.6-bin.tar.gz
[root@master local]# mv apache-maven-3.8.6 maven
[root@master local]# chown -R hadoop:hadoop maven
[root@master local]# ls -ld maven
drwxr-xr-x. 6 hadoop hadoop 99 11月  2 08:33 maven
```

图8-27　设置目录权限

2. 编写程序

切换到hadoop用户。

```
# su - hadoop
```

（1）创建应用程序代码目录

在hadoop用户Home目录创建一个文件夹sparkapp作为应用程序根目录，并在sparkapp文件夹创建图8-28所示的目录结构。

```
$ cd ~
$ mkdir -p ./sparkapp/src/main/scala
```

> **笔记**
>
> Maven是一个项目管理工具，它包含了一个项目对象模型（Project Object Model），一组标准集合，一个项目生命周期（Project Lifecycle），一个依赖管理系统（Dependency Management System），和用来运行定义在生命周期阶段（phase）中插件（plugin）目标（goal）的逻辑。

> **说明**
>
> 随着版本更新，当前版本号的下载地址可能发生变化，请读者自行到官网获取相应下载地址。

> **说明**
>
> 如果通过wget指令获取安装包太慢，可以使用本书提供的安装包。使用scp指令从Windows传输文件至Linux中。

```
[hadoop@localhost ~]$ cd ~
[hadoop@localhost ~]$ mkdir -p sparkapp/src/main/scala
[hadoop@localhost ~]$ find sparkapp/
sparkapp/
sparkapp/src
sparkapp/src/main
sparkapp/src/main/scala
```

图8-28 创建应用程序代码目录

（2）创建代码文件

使用vi在"~/sparkapp/src/main/scala"下建立一个名为SimpleApp.scala的Scala代码文件。

```
$ vi ./sparkapp/src/main/scala/SimpleApp.scala
```

然后在SimpleApp.scala代码文件中输入以下代码，如图8-29所示。

```
/* SimpleApp.scala */
import org.apache.spark.SparkContext
import org.apache.spark.SparkContext._
import org.apache.spark.SparkConf

object SimpleApp {
    def main(args: Array[String]) {
        val logFile = "file:///usr/local/spark/README.md"
        val conf = new SparkConf().setAppName("Simple Application")
//创建SparkConf对象，设置AppName
        val sc = new SparkContext(conf)
        val logData = sc.textFile(logFile, 2).cache()//读取Linux本地文件系统文件/usr/local/spark/README.md创建RDD（分区数为2），保存在缓存中
        val numAs = logData.filter(line => line.contains("a")).count()//统计logData中包含"a"的行数
        val numBs = logData.filter(line => line.contains("b")).count()
        println("Lines with a: %s, Lines with b: %s".format(numAs, numBs))
// 统计logData中包含"a"的行数
    }
}
```

 说明

这段代码的功能是计算/usr/local/spark/README.md文件中包含a的行数和包含b的行数，然后把统计结果打印出来。

```
[hadoop@localhost ~]$ vi sparkapp/src/main/scala/SimpleApp.scala
[hadoop@localhost ~]$ cat sparkapp/src/main/scala/SimpleApp.scala
import org.apache.spark.SparkContext
import org.apache.spark.SparkContext._
import org.apache.spark.SparkConf

object SimpleApp {
    def main(args: Array[String]) {
        val logFile = "file:///usr/local/spark/README.md"
        val conf = new SparkConf().setAppName("Simple Application")
        val sc = new SparkContext(conf)
        val logData = sc.textFile(logFile, 2).cache()
        val numAs = logData.filter(line => line.contains("a")).count()
        val numBs = logData.filter(line => line.contains("b")).count()
        println("Lines with a: %s, Lines with b: %s".format(numAs, numBs))
    }
}
```

图8-29 创建代码文件

3. 打包程序

（1）新建文件pom.xml

使用vi编辑器在"~/sparkapp"目录中新建文件pom.xml。

```
$ cd ~
$ vi ./sparkapp/pom.xml
```

在pom.xml文件中添加如下内容，用来声明该独立应用程序的信息以及与Spark的依赖关系，如图8-30所示。

```xml
<project>
    <groupId>cn.edu.zjitc</groupId>
    <artifactId>simple-project</artifactId>
    <modelVersion>4.0.0</modelVersion>
    <name>Simple Project</name>
    <packaging>jar</packaging>
    <version>1.0</version>
    <repositories>
        <repository>
            <id>jboss</id>
            <name>JBoss Repository</name>
            <url>http://repository.jboss.com/maven2/</url>
        </repository>
    </repositories>
    <dependencies>
        <dependency> <!-- Spark dependency -->
            <groupId>org.apache.spark</groupId>
            <artifactId>spark-core_2.11</artifactId>
            <version>2.1.0</version>
        </dependency>
    </dependencies>

<build>
<sourceDirectory>src/main/scala</sourceDirectory>
    <plugins>
        <plugin>
            <groupId>org.scala-tools</groupId>
            <artifactId>maven-scala-plugin</artifactId>
            <executions>
                <execution>
                    <goals>
                        <goal>compile</goal>
                    </goals>
                </execution>
            </executions>
            <configuration>
                <scalaVersion>2.11.8</scalaVersion>
                <args>
                    <arg>-target:jvm-1.5</arg>
                </args>
            </configuration>
        </plugin>
    </plugins>
</build>
</project>
```

笔记

pom.xml 解释见表 8-4。

表8-4 pom.xml解释

内容	解释
groupId	公司或者组织的唯一标志
artifactId	本项目的唯一ID
modelVersion	模型版本。maven 2.0必须这样写，目前是maven 2.0唯一支持的版本
packaging	打包的机制，如pom、jar、maven-plugin、ejb、war、ear、rar、par，默认为jar
version	本项目目前所处的版本号
repositories	配置maven下载jar的中央仓库
dependencies	配置项目相关依赖
build	编辑工程构建过程中需要配置的资源信息和构建插件信息

```
[hadoop@localhost ~]$ vi sparkapp/pom.xml
[hadoop@localhost ~]$ cat sparkapp/pom.xml
<project>
    <groupId>cn.edu.xmu</groupId>
    <artifactId>simple-project</artifactId>
    <modelVersion>4.0.0</modelVersion>
    <name>Simple Project</name>
    <packaging>jar</packaging>
    <version>1.0</version>
    <repositories>
        <repository>
            <id>jboss</id>
            <name>JBoss Repository</name>
            <url>http://repository.jboss.com/maven2/</url>
        </repository>
    </repositories>
    <dependencies>
        <dependency> <!-- Spark dependency -->
```

图8-30 新建pom.xml

（2）检查文件结构

为了保证Maven能够正常运行，先执行如下命令检查整个应用程序的文件结构。

```
$ cd ~/sparkapp
$ find
```

文件结构如图8-31所示。

```
[hadoop@localhost ~]$ cd sparkapp/
[hadoop@localhost sparkapp]$ find .
.
./src
./src/main
./src/main/scala
./src/main/scala/SimpleApp.scala
./pom.xml
```

图8-31 应用程序文件结构

注意

一定要把应用程序路径设置为当前目录后再进行打包。

注意

第一次打包的过程中，系统会从在线的Maven库中下载相关文件，请保持网络畅通，并耐心等待一定时间。

（3）打包应用程序成JAR包

将整个应用程序打包成JAR包，如图8-32所示。

```
$ cd ~/sparkapp
$ /usr/local/maven/bin/mvn package
```

```
[INFO] Building jar: /home/hadoop/sparkapp/target/simple-project-1.0.jar
[INFO] ------------------------------------------------------------------
[INFO] BUILD SUCCESS
[INFO] ------------------------------------------------------------------
[INFO] Total time:  02:38 min
[INFO] Finished at: 2022-08-28T08:05:39-07:00
[INFO] ------------------------------------------------------------------
```

图8-32 应用程序打包

生成的应用程序JAR包位置为："/home/hadoop/sparkapp/target/simple-project-1.0.jar"，如图8-33所示。

```
[hadoop@master sparkapp]$ ls /home/hadoop/sparkapp/target/
classes            maven-archiver    simple-project-1.0.jar
classes.timestamp  maven-status
```

图8-33 应用程序JAR包位置

4. 通过spark-submit运行程序

对于前面Maven打包得到的应用程序JAR包，可以通过spark-submit提交到Spark中运行，如图8-34所示。

说明

spark-submit 是 /usr/local/spark/bin/ 路径下的可执行文件，由于设置了 PATH 变量，因此此处直接输入可执行文件名即可。

```
$ spark-submit --class "SimpleApp" ~/sparkapp/target/simple-project-1.0.jar
```

图8-34　spark-submit命令执行效果1

上面命令执行后会输出太多信息，可以使用下面命令查看想要的结果：

```
$ spark-submit --class "SimpleApp" ~/sparkapp/target/simple-project-1.0.jar 2>&1 | grep "Lines with a:"
```

执行结果如图8-35所示。

图8-35　spark-submit命令执行效果2

拓展学习

1. Spark运行架构

（1）Spark的相关基础概念

Spark的相关基础概念见表8-5。

扫码看视频

表8-5　Spark的相关基础概念

名　　称	解　　释
RDD	是Resillient Distributed Dataset（弹性分布式数据集）的简称，是分布式内存的一个抽象概念，提供了一种高度受限的共享内存模型
DAG	是Directed Acyclic Graph（有向无环图）的简称，反映RDD之间的依赖关系
Executor	是运行在工作节点（WorkerNode）的一个进程，负责运行Task
Application	用户编写的Spark应用程序
Task	运行在Executor上的工作单元
Job	一个Job包含多个RDD及作用于相应RDD上的各种操作
Stage	是Job的基本调度单位，一个Job会分为多组Task，每组Task被称为Stage，或者也被称为TaskSet，代表了一组关联的、相互之间没有Shuffle依赖关系的任务组成的任务集

一个Application由一个Driver和若干个Job构成，一个Job由多个Stage构成，一个Stage由多个没有Shuffle关系的Task组成。当执行一个Application时，Driver会向集群管理器申请资源，启动Executor，并向Executor发送应用程序代码和文件，然后在Executor上执行Task，运行结束后，执行结果会返回给Driver，或者写到HDFS或者其他数据库中。

（2）架构设计

如图8-36所示，Spark运行架构包括集群资源管理器（Cluster Manager）、运行作业任务的工作节点（Worker Node）、每个应用的任务控制节点（Driver）和每个工作节点上负责具体任务的执行进程（Executor）。集群资源管理器可以是自带的资源管理器也可以是YARN或Mesos等资源管理框架。

与Hadoop MapReduce计算框架相比，Spark所采用的Executor有两个优点：一是利用多线程来执行具体的任务，减少任务的启动开销；二是Executor中有一个BlockManager存储模块，会将内存和磁盘共同作为存储设备，有效减少IO开销。

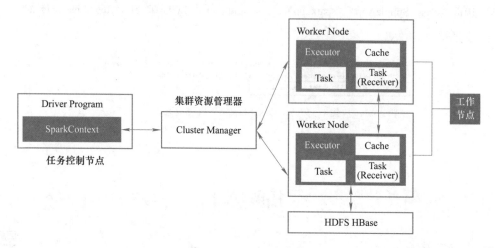

图8-36　Spark运行架构

（3）Spark运行基本流程

Spark运行基本流程如图8-37所示。

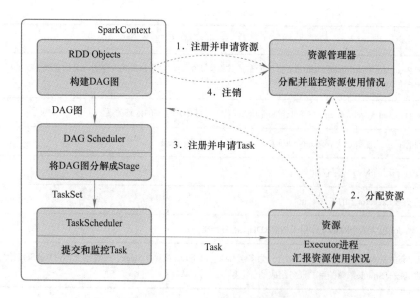

图8-37　Spark运行基本流程

1）首先为应用构建起基本的运行环境，即由Driver创建一个SparkContext，进行资源的申请、任务的分配和监控。

2）资源管理器为Executor分配资源，并启动Executor进程。

3）SparkContext根据RDD的依赖关系构建DAG图，DAG图提交给DAGScheduler解析成Stage，然后把一个个TaskSet提交给底层调度器TaskScheduler处理；Executor向SparkContext申请Task，TaskScheduler将Task发放给Executor运行，并提供应用程序代码。

4）Task在Executor上运行，把执行结果反馈给TaskScheduler，然后反馈给DAGScheduler，运行完毕后写入数据并释放所有资源。

2. RDD的设计与运行原理

（1）RDD概念

一个RDD就是一个分布式对象集合，本质上是一个只读的分区记录集合，每个RDD可分成多个分区，每个分区就是一个数据集片段，并且一个RDD的不同分区可以被保存到集群中的不同节点上，从而在集群中的不同节点上进行并行计算。

RDD提供了一种高度受限的共享内存模型，即RDD是只读的记录分区的集合，不能直接修改，只能基于稳定的物理存储中的数据集创建RDD，或者通过在其他RDD上执行确定的转换操作（如map、join和group by）而创建得到新的RDD。

（2）RDD的运行过程

RDD的运行流程如图8-38所示，具体流程为：

1）创建RDD对象。

2）SparkContext负责计算RDD之间的依赖关系，构建DAG。

3）DAGScheduler负责把DAG图分解成多个Stage，每个Stage中包含了多个Task，每个Task会被TaskScheduler分发给各个WorkerNode上的Executor去执行。

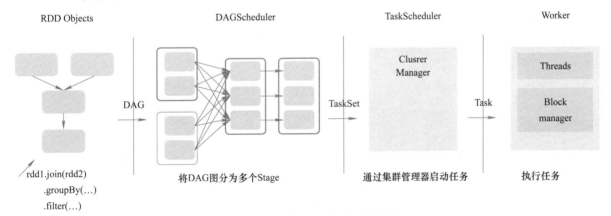

图8-38　RDD在Spark中的运行过程

项目小结

本项目围绕"部署与使用Spark"典型工作任务，以初入职场的大数据运维工程师视角，讲解了Spark、Spark的生态系统、Spark与Hadoop、Spark的优势、Spark的部署模式、Spark Shell基本操作等基本概念，展开四个任务，学习如何部署与使用Spark，部署部分任务的横向对比如图8-39所示，引导大数据运维工程师新人逐步掌握部署与使用Spark的相关能力。在完成任务后，开展扩展知识学习，包括Spark的运行架构、RDD的设计与运行原理。

	任务1：部署单机模式Spark	任务2：部署Spark集群
安装Spark	下载Spark安装文件 解压安装文件 配置目录权限 配置PATH	下载Spark安装文件 解压安装文件 配置目录权限 配置PATH
配置Spark	配置spark-env.sh文件	配置spark-env.sh文件 配置workers文件
同步配置		同步spark文件夹到slave节点 配置目录权限
运行Spark	运行SparkPi	启动Spark集群

图8-39　Spark部署任务横向对比总结

实战强化

通过Spark的RDD编程，实现词频统计的功能。

提示：对文件"/usr/local/spark/README.md"进行词频统计。

参 考 文 献

[1] TOM W. Hadoop 权威指南：大数据的存储与分析[M]. 王海，华东，刘喻，等译. 北京：清华大学出版社，2017.

[2] 陆嘉恒. Hadoop 实战[M]. 2版. 北京：机械工业出版社，2012.

[3] DIMIDUK N，KHURANA A. HBase 实战[M]. 谢磊，译. 北京：人民邮电出版社，2013.

[4] CHAMBERS B，ZAHARI M. Spark 权威指南[M]. 张岩峰，王方京，译. 北京：中国电力出版社，2020.

[5] 林子雨. 大数据技术原理与应用：概念、存储、处理、分析与应用[M]. 3版. 北京：人民邮电出版社，2021.

[6] 新华三技术有限公司. 大数据平台运维：初级[M]. 北京：电子工业出版社，2021.

[7] 新华三技术有限公司. 大数据平台运维：中级[M]. 北京：电子工业出版社，2021.

[8] 黑马程序员. Hadoop 大数据技术原理与应用[M]. 北京：清华大学出版社，2019.